IoT-Enabled Healthcare Systems

Applications, Benefits, and Challenges

AAP Series on Digital Signal Processing, Computer Vision and Image Processing

IoT-Enabled Healthcare Systems

Applications, Benefits, and Challenges

Edited by

Yatindra Kumar, PhD

Shashi Kant Verma, PhD

Archana Singh, MBBS, MD

Krishna Kumar, PhD

Manoj Gupta, PhD

AAP | APPLE ACADEMIC PRESS

First edition published 2025

Apple Academic Press Inc.
1265 Goldenrod Circle, NE,
Palm Bay, FL 32905 USA

760 Laurentian Drive, Unit 19,
Burlington, ON L7N 0A4, CANADA

CRC Press
2385 NW Executive Center Drive,
Suite 320, Boca Raton FL 33431

4 Park Square, Milton Park,
Abingdon, Oxon, OX14 4RN UK

Library and Archives Canada Cataloguing in Publication

Title: IoT-enabled health care systems : applications, benefits, and challenges / edited by: Yatindra Kumar, Shashi Kant Verma, Archana Singh, Krishna Kumar, Manoj Gupta.
Names: Kumar, Yatindra (Lecturer in electrical engineering), editor. | Verma, Shashi Kant, editor. | Singh, Archana (Lecturer in radiotherapy), editor. | Kumar, Krishna (Engineer), editor | Gupta, Manoj (Lecturer in computer science and engineering), editor.
Series: AAP series on digital signal processing, computer vision and image processing.
Description: First edition. | Series statement: AAP series on digital signal processing, computer vision and image processing | Includes bibliographical references and index.
Identifiers: Canadiana (print) 20240447719 | Canadiana (ebook) 20240447778 | ISBN 9781774916841 (hardcover) | ISBN 9781774916858 (softcover) | ISBN 9781003505235 (ebook)
Subjects: LCSH: Medical care—Data processing. | LCSH: Medical informatics. | LCSH: Internet of things.
Classification: LCC R858 .I58 2025 | DDC 610.285—dc23

Library of Congress Cataloging-in-Publication Data

CIP data on file with US Library of Congress

ISBN: 978-1-77491-684-1 (hbk)
ISBN: 978-1-77491-685-8 (pbk)
ISBN: 978-1-00350-523-5 (ebk)

AAP SERIES ON DIGITAL SIGNAL PROCESSING, COMPUTER VISION AND IMAGE PROCESSING

BOOK SERIES EDITORS:

Dr. Manoj Gupta, PhD
Associate Professor, Department of Electronics and Communication Engineering, JECRC University, Jaipur (Rajasthan), India
Email: manojgupta35@yahoo.co.in

Dr. Pradeep Kumar, PhD
Discipline of Electrical, Electronic and Computer Engineering,
Howard College Campus, University of KwaZulu-Natal,
Durban-4041, South Africa
Email: pkumar_123@yahoo.com, kumarp@ukzn.ac.za

Brief description of the Book Series:

Digital signal processing, computer vision and image processing as a whole is considered to be one of the most rapidly evolving areas of research and technology today with growing applications in almost all disciplines of engineering. Medical imaging, computer vision, healthcare, medical applications, remote sensing, agriculture, robotics, communication systems and space exploration are some of the applications of digital signal processing, computer vision and image processing, to name a few. The present day curriculum covers many aspects of digital signal processing, computer vision and image processing, addressing the theoretical aspects in particular. This book series is intended to supplement the theoretical knowledge with special emphasis on the practical side. The series content has been specifically chosen to give a thorough understanding of the fundamental aspects to advanced applications of digital signal processing, computer vision and image processing.

The series is open to monograph. handbooks, authored books, edited volumes, and conference proceedings. This series aim to embrace all aspects, sub fields and new challenges in the followings research domains (related topics) but are not limited to:

- **Image and Video Processing:** Image filtering, restoration and enhancement, image segmentation, video segmentation and tracking, morphological processing, feature extraction and analysis, interpolation and super-resolution, motion detection and estimation, computer vision, pattern recognition, content-based image retrieval, image/signal computations and services, features and models for image/signals, machine learning based image and signal processing, data mining techniques, imaging algebra, mathematical morphology, probabilistic, statistical, optimization, approximation theory, models in imaging science, video signal processing, visualization, watermarking, video surveillance, video compression and streaming, video analysis and event recognition, biometrics , medical image analysis, artificial intelligence and related areas.
- **Signal Processing:** Filters theory, spectral analysis, time-frequency and time-scale representation, EEG/ECG signals, FIR/IIR and adaptive filters, statistical signal processing, filtering, detection and estimation, nonlinear signal processing, radar, antennas, telecommunications systems, acoustics. Signal processing theory and methods, high signal processing: integrating 5G and IoT with satellite networks, hands-free speech communication and microphone arrays, wearable sensor signal processing, architecture and frameworks, audio/speech processing and coding, watermarking, data mining techniques, statistical and optical signal processing, communication signal processing, DSP system and embedded systems, multimedia processing, artificial intelligence , IoT, cyber physical systems and related areas.
- **Computer Vision:** Algorithms, feature extraction and pattern recognition, ML and deep learning in vision, CBIR, object and face recognition, AR/VR, object detection and localization, 3D object extraction, tracking and visual navigation, cognitive and biological inspired vision, artificial intelligence, machine learning, ubiquitous computing and related areas.

- **Applications Areas:** Biometric, bioinformatics and biomedical imaging, medial images and applications, healthcare applications, agriculture applications, augmented and mixed reality, mental health and cognitive stimulations, security and video-surveillance, quality control and inspection, archaeology, remote sensing, embedded systems and applications, automated vehicles, speech and music processing, robotics, rehabilitation, occupational therapy and telemedicine, artificial intelligence and machine learning/deep learning based applications, cyber physical systems, Internet of Things (IoT), Industry 4.0, Medicine 5.0 and other related applications.

BOOKS IN THE SERIES:

Artificial Intelligence and Machine Learning Techniques in Image Processing and Computer Vision
Editors: Karm Veer Arya, PhD, Ciro Rodriguez Rodrigues, PhD, Saurabh Singh, PhD, and Abhishek Singhal, PhD

Computational Imaging and Analytics in Biomedical Engineering: Algorithms and Applications
Editors: T. R. Ganesh Babu, PhD, U. Saravanakumar, PhD, and Balachandra Pattanaik, PhD

IoT-Enabled Healthcare Systems: Applications, Benefits, and Challenges
Editors: Yatindra Kumar, PhD, Shashi Kant Verma, Archana Singh, Krishna Kumar, PhD, and Manoj Gupta, PhD

ABOUT THE EDITORS

Yatindra Kumar, PhD
Professor, Department of Electrical Engineering, Govind Bhallabh Pant Institute of Engineering & Technology (formerly G. B. Pant Engineering College), Pauri Garhwal, Uttarakhand, India

Yatindra Kumar, PhD, is currently working as a Professor, Department of Electrical Engineering, Govind Bhallabh Pant Institute of Engineering & Technology (formerly known as G. B. Pant Engineering College") (an autonomous institute of Uttarakhand Government), Pauri Garhwal, Uttarakhand, India. He has more than 18 years of teaching, administrative, and research experience. His current research interests include signal processing and image processing, AI, machine learning, and IoT. Dr. Kumar has published over 30 research papers in reputed international journals and conferences, including IEEE, Springer, Elsevier, Wiley & Sons, Taylor & Francis, Inderscience, etc. He did his BE in Electronics & Communication Engineering, MTech, in Control and Instrumentation from MNNIT, Allahabad U.P., and PhD in Biomedical Signal Processing from IIT Roorkee, India.

Shashi Kant Verma, PhD
Associate Professor, Department of Computer Science; Engineering, Govind Bhallabh Pant Institute of Engineering & Technology (formerly G. B. Pant Engineering College), Pauri Garhwal, Uttarakhand, India

Shashi Kant Verma, PhD, is currently working as an Associate Professor in the Department of Computer Science Engineering at the Govind Bhallabh Pant Institute of Engineering & Technology (formerly known as G. B. Pant Engineering College) (an autonomous institute of Uttarakhand Government), Pauri Garhwal, Uttarakhand, India. He has more than 18 years of teaching, administrative, and research experience. His current research interest includes embedded systems, wireless sensor networks, reliable computing, fault-tolerant systems, computer organization, architecture, and distributed systems. Dr. Verma has published over 50 research papers in reputed international journals and conferences, including IEEE, Springer, Wiley & Sons, Taylor & Francis, Inderscience, Hindawi, etc. He did his BE in Electronics & Communication Engineering, MTech, PhD in Computer Science & Engineering.

Archana Singh, MBBS, MD
Assistant Professor, J. K. Cancer Institute, GSV Medical College, Kanpur, India

Archana Singh, MBBS, MD (Radiation Oncology), holds degrees from Sanjay Gandhi Postgraduate Institute of Medical Sciences, Lucknow, India, and is presently working as an Assistant Professor in the Radiotherapy Department of the J K Cancer Institute (GSVM Medical College) Kanpur. She earlier worked as a Senior Resident at AIIMS, New Delhi. Her field of interest is breast cancer and gynecological malignancies. She has a number of publications in national as well as international journals.

Krishna Kumar, PhD
Research and Development Engineer, Uttarakhand Jal Vidyut Nigam Ltd. Dehradun, (UJVNL) (A Govt. of Uttarakhand Enterprises), India

Krishna Kumar, PhD, is currently a Researcher in the Department of Water Resources Engineering and the Center for Advanced Middle Eastern Studies at Lund University, Sweden. Previously, he held the position of Senior Engineer Research at the Rolls-Royce @ NTU Corporate Lab, Singapore. He also served as a postdoctoral research fellow at the School of Electrical and Electronics Engineering at the Rolls-Royce Corporate Lab, Nanyang Technological University (NTU), Singapore. He was an Assistant Engineer at UJVN Ltd. (A State Government PSU of Uttarakhand). His expertise includes the design, installation, commissioning, testing, and operation and maintenance of hydro, solar, and wind power plants. Dr. Kumar has also served as an Assistant Professor/Lecturer on a term appointment basis at Kumaon Engineering College, Dwarahat, Uttarakhand; Bharat Institute of Technology, Meerut; and Uttaranchal Institute of Technology, Dehradun, India. Dr. Kumar has published numerous research articles in reputed journals and has also authored/edited many books with publishers such as Elsevier, Wiley, Taylor & Francis, Springer, and River Press. His current research interests include renewable energy, multiphase machines, electrical equipment health monitoring systems, AI/ML, and data science.

Dr. Kumar earned his bachelor's in Electronics and Communication Engineering at Govind Ballabh Pant Engineering College, Pauri Garhwal, Uttarakhand, India. He pursued his master's in digital systems at the Motilal Nehru National Institute of Technology, Allahabad, India, and received his PhD degree at the Indian Institute of Technology Roorkee, India.

Manoj Gupta, PhD
Professor, School of Computer Science and Engineering (SCOPE), VIT-AP University, Amravati, Andhra Pradesh, India

Manoj Gupta, PhD, is presently working as a Professor at the School of Computer Science and Engineering (SCOPE), VIT-AP University, Amravati (Andhra Pradesh), India. He has an overall experience of more than 18 years. He has published over 60 research papers in international journals and national and international conferences as well as book chapters. He has two patent grants and has published five patents and have two Indian copyrights to his credit. He is the Editor-in-Chief of a number of book series, including the book series *AAP Book Series on Digital Signal Processing, Computer Vision and Image Processing*. His research interests are in antenna and wireless communications, biomedical image and signal processing, soft computing, computational intelligence, biomedical engineering, artificial intelligence, and neuroscience. He is a member of many professional bodies. He has also served as technical program committee member and reviewer for various international conferences. He also serves as editor, associate editor, and reviewer for many international journals. His name has been listed in *Marquis Who's Who in Science and Engineering® USA and Marquis Who's Who in the World® USA*. He was awarded his PhD in Computer Science from the University of Rajasthan, Jaipur, India. He received his MTech in Digital Signal Processing from H.N.B Garhwal Central University, Uttarakhand, India, and BTech degree in Electronics and Communication Engineering from the Institute of Engineering & Technology, M.J.P. Rohilkhand University Campus, Bareilly (UP), India.

CONTENTS

CONTRIBUTORS

Kriti Ahuja
Department of Allied Sciences, School of Health Science and Technology, University of Petroleum and Energy Studies, Dehradun, Uttarakhand, India

Y. S. Chauhan
Department of Computer Applications, G. B. Pant Institute of Engineering and Technology, Pauri Garhwal, Uttarakhand, India

Siddarth Ghensal
Department of Computer Applications, G. B. Pant Institute of Engineering and Technology, Pauri Garhwal, Uttarakhand, India

Nishu Goyal
Department of Allied Sciences, School of Health Science and Technology, University of Petroleum and Energy Studies, Dehradun, Uttarakhand, India

Manoj Gupta
Department of Electrical Engineering, Faculty of Engineering, Chulalongkorn University, Bangkok, Thailand

Maya Datt Joshi
Department of Biotechnology, GLA University, Mathura, Uttar Pradesh, India

Ila Kaushik
Krishna Institute of Engineering and Technology, Ghaziabad, Uttar Pradesh, India

Krishna Kumar
Department of Hydro and Renewable Energy, Indian Institute of Technology, Roorkee, Uttarakhand, India

Narendra Kumar
School of Computing, DIT University Dehradun, Dehradun, Uttarakhand, India

Surabhi Lingwal
Department of Computer Science and Engineering, G. B. Pant Institute of Engineering and Technology, Pauri Garhwal, India

Banit Negi
Department. of Computer Science and Engineering, G. B. Pant Institute of Engineering and Technology, Pauri Garhwal, India

Anjum Sheikh Qureshi
Rajiv Gandhi College of Engineering Research and Technology, Chandrapur, India

Indradeo Ram
Department of Computer Science and Engineering, BIT Mesra, Ranchi, Jharkhand, India

J. S. Rauthan
Department of Computer Science and Engineering, G. B. Pant Institute of Engineering and Technology, Pauri-Garhwal, India

Mhd. Waseem Raza
J K Cancer Institute, Kanpur, Uttar Pradesh, India

Minakshi Sehrawat
The NorthCap University, Gurugram, Haryana, India

Deergha Sharma
The NorthCap University, Gurugram, Haryana, India

Nikhil Sharma
Delhi Technological University, Delhi, India

Amit Kumar Singh
Biomedical Engineering Department, VSB Engineering College, Karur, Tamil Nadu, India

Archana Singh
J K Cancer Institute, Kanpur, Uttar Pradesh, India

Deepak Kumar Verma
Greater Noida Institute of Technology, Greater Noida, India

Shashikant Verma
G.B. Pant Institute of Engineering and Technology, Pauri Garhwal, Uttarakhand, India

Shubham Verma
Tata Consultancy Services, India

S. K. Verma
Department of Computer Science and Engineering, G. B. Pant Institute of Engineering and Technology, Pauri-Garhwal, India

PREFACE

According to the World Health Organization's (WHO) guidelines, the doctor–population ratio should be 1:1000, but several countries have struggled to achieve it. Therefore, a system is needed to fill this ratio gap with the aid of technology. A healthcare system that uses IoT technology is reliable and secure in allowing the best use of the available resources. It also helps in improving contact between the community and public healthcare. Patient contacts with physicians prior to the IoT were limited to visits and tele- and text messages. There was no way physicians or hospitals could constantly track the health of patients and make recommendations accordingly.

The IoT-enabled tools have made remote monitoring in the healthcare sector possible, unleashing the ability to keep patients safe and secure and inspiring doctors to deliver excellent care. This has also improved patient interest and satisfaction as it has become easier and more effective to communicate with doctors. In addition, video monitoring of the well-being of patients tends to minimize the duration of hospital stays and avoids readmissions.

IoT also has a significant impact on significantly lowering healthcare costs and improving treatment outcomes. IoT is undeniably changing the healthcare industry by redefining the system space and the relationship between people in providing healthcare solutions. IoT has healthcare applications that support patients, families, physicians, hospitals, and insurance companies. Developments in technology influence every aspect of the everyday life opportunities of all living beings. Research and development in different industries such as communication, sensors, internet technology, and so on, also influence the healthcare industry in several cases. Hospitals and clinics will enhance access to patient care while increasing the quality of treatment and reducing operational costs by implementing IoT-based systems. An IoT-based system's basic building blocks are sensors that collect patient data, internet gateways to relay that data, and cloud storage to store and process this data.

The smart cloud platform is often used to analyze these data in order to produce valuable knowledge for physicians and medical personnel, also requiring the integration of an ML algorithm, AI, and IoT technologies in healthcare systems to derive useful information from the data collected. The final stage includes designing online and mobile apps that can be used by medical personnel to decide on the next course of action. Patients are also given mobile devices, apps, and even wearable devices to monitor their health in patient-centered IoT systems.

If this program is in place, one of its primary uses is to build a better method for treating chronic diseases. Many devices can now be used to track chronic patients, such as asthma, diabetes, hypertension, and chronic obstructive pulmonary disease. The off-the-shelf and advanced wearable can help patients and physicians track operations. In addition, helpful in-depth and realistic comprehension tests and patient-reported results are the routine evaluations of the app, which demonstrate that cognitive tests administered by an app can be rigorous as well as accurate.

IoT in healthcare, however, is not without challenges. IoT-enabled connected devices collect massive quantities of data, including confidential information, which gives rise to data security concerns. Implementation of sufficient precautions is essential.

IoT explores new aspects of patient care through real-time clinical monitoring and access to health data from patients. This data is a goldmine for healthcare providers to enhance the safety and experiences of patients while generating incentives for growth and enhancing healthcare operations. Getting prepared to harness the digital power in that connected world will prove to be the differentiator. The proliferation of IoT-specific healthcare products opens up massive opportunities, and the vast amount of data produced by these connected devices holds the potential for transforming healthcare systems.

CHAPTER 1

EXPERT SYSTEMS FOR HEALTHCARE

NARENDRA KUMAR[1], KRISHNA KUMAR[2], and INDRADEO RAM[3]

[1]*School of Computing, DIT University Dehradun, Dehradun, Uttarakhand, India*

[2]*Department of Hydro and Renewable Energy, Indian Institute of Technology, Roorkee, India*

[3]*Department of Computer Science and Engineering, BIT Mesra, Ranchi, Jharkhand, India*

ABSTRACT

There has long been a consensus that healthcare systems play a critical role in enhancing human well-being and contributing to a healthy society. Efforts should be made to promote public awareness and execute necessary measures to improve healthcare. Because of this, a computer-based system is created to allow individuals to self-diagnose their health state based on specified symptoms. This approach should be stressed so that individuals may appropriately use the knowledge as a reference to live a better life. Web-based Community Center for Healthcare Diagnosis is designed using an expert system method. Treatment and prevention of illness may be determined using expert system reasoning techniques, which are implemented into the system. At the moment, arthritis, thalassemia, and pneumococcal are included in the list of illnesses. The knowledge-based system manages sets of rules and facts. Web-based technology is used to spread information to users so that they may utilize it to their advantage. By using self-diagnosis for early illness identification, this approach will

IoT-Enabled Healthcare Systems: Applications, Benefits, and Challenges
Yatindra Kumar, Shashi Kant Verma, Archana Singh, Krishna Kumar, & Manoj Gupta (Eds.)

assist those who want to raise their health awareness and seek professional information about diseases.

1.1 INTRODUCTION

The diagnosis in medicine is the process of figuring out what illness or condition is causing a person's signs and symptoms. An early diagnosis of an illness is critical for a person to begin treatment. Healthcare facilities have medical professionals on staff that can aid with medical diagnostics. In the real world, anybody may get a medical diagnosis from a human expert at any hospital or clinic. Due to time constraints, medical check-up prices, or a person's personal attitude, some persons are unable to undertake specialized medical examinations. Regardless, keeping tabs on one's health is an absolute must for everyone. Many sophisticated strategies are used to simulate medical diagnosis in a computer-based system in this era of information and Internet technologies. As a result of these strategies, a computer system may mimic the actions of an expert inferring a judgment from supplied information. A self-diagnosis system based on intelligent information and available through the Internet enables anyone to do their own tests at their own leisure. Intelligent knowledge-based system structures enable human specialists' knowledge in a computer system to be preserved Practicing self-diagnosis is an essential element of personal healthcare. An optimistic outlook on life may boost one's self-confidence and self-esteem.[1,3] We all need to practice self-care as individuals and as a community. Self-care and rule-base management might both benefit from expert systems that produce a person's state based on symptoms. Even more important, the preservation of human expertise via the rule set knowledge base system is critical. As a result, experts' expertise may be shared with the greater community for the benefit of society. It is the goal of expert system research to make expert information available to decision makers anytime they are in a hurry to find a solution to a dilemma. Consequently, the goal of this project was to construct an integrated knowledge-based expert system for numerous diagnostic applications. With the help of experts, this system will help raise public awareness of sickness in an alternative way. At the moment, this system includes three expert system applications. More expert systems for Healthcare applications should be available in the near future, according to expectations. In other words, this system may be used to manage several expert systems across a wide range of medical specialties and settings. Physicians and patients may use this system to get an early diagnosis and treatment recommendations thanks to the information it gives. Using the medical expert system, you can get the same information that a medical expert would. Using a single platform, the multi-disease expert system can make accurate diagnoses. As a result, medical professionals may disseminate information about illnesses to the general population in an effort to raise awareness and improve access to Healthcare. Human expert reasoning is modified and handled dynamically in the integrated knowledge-based expert system, which has two major benefits. Second, a single platform is used to preserve and disseminate specialized information.

1.2 RELATED WORK

Eleven urban hospitals were evaluated using a fuzzy stochastic DEA approach based on the slacks-based measure of efficiency (SBM) to determine their performance. A decision-making trial and evaluation laboratory (DEMATEL) and an analytical network process (ANP) integrated algorithm is used to discover the most important and effective criterion due to the diversity of the first established criteria. Because of using the method, total inpatient visits are the most significant criteria. Finally, the suggested model is applied using multiple degrees of importance, and the performance of each unit is determined under unknown circumstances.[2] Initially, most of the data and photographs were gathered from patients, the Internet, and specialized physicians, and then stored in SQL Server by the system developers. There were a variety of data fields in this database to store information about each patient, including their medical history, diagnosis, symptoms, and reactions. Several methods are used to refine the photos once they have been gathered.[4–14] There has been a lot of interest in the use of machine learning (ML) in Healthcare. Medical care may benefit from the use of ML methods when massive data and improved computing capacity are combined. This sort of supervised learning may be used to predict labeled data using algorithms like linear or logistic regression, support vector machine, decision tree LASSO regression and Naive Bayes classifier. When training datasets without any outcome information, unsupervised ML models may find patterns in the data that would otherwise be missed by supervised models. Detection of fraud or anomalies may be done using these models. Many clinical decision support systems may be developed using ML. For example, ML may be used in public health to identify and forecast vulnerable groups, as well as build focused public health treatments for these populations. The medical curriculum should include principles related to ML in order for health professionals to properly guide and analyze research in this field.[15] Healthcare providers, like cancer survivors, have differing attitudes and methods to self-management and support for self-management in this group. That's why it is helpful to have people in self-management support positions that are both instructive and collaborative. Patients and coworkers benefit from healthcare practitioners who are self-aware and communicate these opinions openly. As part of self-management training, students should be taught how to reflect on their own behavior and how it relates to professional treatment.[16] The systematic literature review was carried out using a hermeneutic method and a recommended reporting items for systematic review and meta-analysis (PRISMA) technique. For AI solution developers, healthcare experts, and policymakers, we present a framework that incorporates five essential issues from AI governance and ethics. All of these concepts are summed up in the acronym SHIFT: Sustainability, Human centeredness, Inclusion, Fairness, and Transparency. This chapter also identifies and discusses potential research directions for future work on ethical use of artificial intelligence in Healthcare.[17] In the face of an increasing number of breast cancer patients who must be cared for with limited Healthcare resources, new digital technologies may help, according to recent research.[18] COVID-19 patients were surveyed from January 25 to February 22, 2021, using a web-based survey of Japanese healthcare facilities. Japan's health officials sent out a survey request to every Healthcare facility. Information on anti-data-leakage techniques and frameworks for effective, sustainable data collecting should be provided in circumstances where COVID-19 reporting is high.[19,20] AI and Big Data analytics are used in the notion of intelligent health (iHealth) in mental Healthcare. Three key features outlined in the literature: self-monitoring, ecological momentary assessment (EMA), and data mining, are highlighted in this chapter in an effort to explain ethical concerns of iHealth. Searches in databases yielded the data needed for this investigation. Authors looked at studies and reviews that provided data on the outcomes of each of the three components. The results of an ethical framing of the findings illustrate the opportunities and difficulties of iHealth. Prevention of mental illness, early detection, tailored therapy, and patient involvement in the process may be possible with the help of a synergistic combination of self-monitoring, event monitoring analysis, and data mining. Users' freedom, privacy, and security of their personal information are all at risk, not to mention the possibility of prejudice.[21] A variety of medical services, such as disease prevention, Healthcare, physical examinations for disease diagnosis and treatment, rehabilitation, chronic disease management, home care, first aid and outpatient attendance and consultations, could benefit from the interaction between virtual and real cloud experts and terminal users (including doctors at the end of their careers and their patients and even their family). It is also important to highlight that the Metaverse in Medicine requires a trustworthy security system, which is the cornerstone for the platform's smooth functioning.[22] Healthcare process modeling activities have shown and documented a number of common difficulties that need to be addressed in this study. As a starting point for the standard modeling of complicated Healthcare elements, Authors provide ready-to-use BPMN process fragments for each difficulty. The findings of a preliminary study on the usage and perceived utility of best practices by younger specialists in health and information technology are also presented. Although BPMN is a valuable tool for capturing Healthcare characteristics, the domain-specific process fragments assist modelers with little or no BPMN or Healthcare domain knowledge to learn more about healthcare processes.[23] A novel way to creating expert systems has been developed that incorporates logical criteria and variable evoking strength in order to increase diagnostic accuracy of these instruments.[24] There are various ML techniques, Internet of things and Image processing have introduced that are used in healthcare system.[25-30]

1.3 EXPERT SYSTEMS IN HEALTHCARE

A precise diagnosis is critical in the Healthcare industry, as a patient's life might be put at risk if even a little mistake is made during treatment or diagnosis. Many different methods are available for building an expert system, all of which rely on well-established algorithms. Data mining techniques are used in most of these strategies. In the process of knowledge discovery, data mining techniques are often used to extract valuable information from vast volumes of data. In this chapter, an expert method for the identification of health problems is provided. Production rules expressed in conventional bi-valued logic and possible facts make up the expert system's knowledge base. Data mining (DM) and case-based reasoning (CBR) are used to construct a model of chronic illness prediction and diagnostic system. An expert service for healthcare services (HES) has been established based on the service being customized by experts to choose the relevant objects for service provision.

Expert systems are a subset or application of artificial intelligence, which is a discipline of computer science. As a result, the term "artificial intelligence" is used to describe a wide range of technologies that enable computers to mimic human intellect. Eoniputer software, when well-crafted, provides decision assistance in the form of diagnostic information and/or therapy or prognosis recommendations. After receiving information (input) on the patient, generally from the patient's physician, the software provides diagnostic, therapeutic, or prognostic guidance. There are distinct differences between expert systems and other types of medical software. These qualities include the expert system's use of clinical reasoning to make a diagnosis or prescribe a course of treatment. Also, in many expert systems, the physician may see the sequence of steps. When it comes to clinical medicine, experts may be able to articulate their views in terms of likelihood. A significant number of facts and rules concerning the sickness or condition in issue is normally agreed upon in order for expert system software to provide reliable responses. Two basic internal medicine textbooks and three specialty textbooks are expected to need 2 million rules. This is based on a conservative estimate. Expert systems can only be utilized with huge, costly computers in the past because to the massive quantities of data they required. When it comes to routine clinical use, expert systems have had limited success "with the advent of more powerful microcomputers and more efficient microcomputer languages" "by the difficulty of creating expert systems" and the "reluctance of physicians to use such programs." The average time it takes to develop an expert system is five years or more "Expert systems with gaps in their functionality are not uncommon. When discussing medical expert systems, it is common to talk about them as though they are two competing methods of medicine. Expert systems may obscure the overall picture of a patient's condition, for example, according to one study "focus on the details of a medical issue" in order. According to some experts, expert systems are not able to replicate a doctor's intuition. "On the other hand, this remark ignores the fact that certain expert systems make use of both principles and data". Maintenance and support of big expert systems may "represent a very prolonged commitment on the part of any organization" since they need continuous update. The creation of modest expert systems has grown more accessible.

1.4 BENEFITS AND SHORTFALLS OF CURRENT EXPERT SYSTEM TECHNOLOGY

The previous three decades of medical expert system research have highlighted the challenge in expressing complicated medical knowledge and building medical expert system applications. There is a shift in medical technology away from the large-scale, research-based systems to the nimbler, practical systems of today. Information management and efficient distribution of medical expertise are becoming more important as more individuals seek out medical information, as well as a means of organizing the massive quantity that is presently accessible in the Healthcare industry. These services may be made available at a lower cost via the use of expert systems because of the following features:

a. The facilitating the development and testing of theories and concepts by professionals.
b. Taking the burden off of specialists. Such duties like regular data analysis or diagnosis of symptoms may be performed by expert systems. Having these chores taken care of frees up the expert's time to focus on the more challenging analysis.
c. Increasing the accessibility of knowledge. A simple and cost-effective method of disseminating expert information, expert systems make it easier for the general public to benefit from it.

Medical information is growing more and more complicated and difficult to comprehend.. An urgent demand exists for instruments that can store and analyze medical knowledge from a variety of sources. Commercial expert system tools range from high-level programming languages to expert system shells and are widely accessible. How can these systems not be employed more often, given their many benefits? Because of this, there are a number of causes.

a. An advanced system: Because of this, it is normally necessary to have a team of programmers to implement shells. Constructing competence in a rule base is tough with shells because of the many issues that arise. This results in exorbitant development costs and timeframes.
b. They have a limited capacity for storing and displaying information. Most expert system shells for personal computers allow you to input data in the form of production rules. The problem is that they are unable to accurately convey medical information. Trying

to translate the expert's intuition into a rule form is a challenge for him or her. Another issue is that when a rule base grows, conflicts between rules arise, making it difficult to update or extend the system.

c. It is difficult to update or contribute to the information and experience that they have. An expert system is sometimes exceedingly difficult and costly to upgrade if knowledge or needs change after it is completed. Medical expertise may be built into a system by an expert who has no prior computer experience, which is one method to make systems more affordable. The expert should be able to input information in a manner that is both familiar to them and easy for the system. In other words, it is not only about offering a more English-like rule structure, but also a more fundamentally natural manner for an expert to break down his intuitive knowledge. In addition to allowing experts to verify their own expertise, it is critical that the system be able to explain its reasoning process. To overcome these practical issues, Authors believe that an entirely new approach to expert system design is necessary. When developing medical expert system tools, the CA TEG05-ES project provides an overview of some of the issues that occur and demonstrates a method for tackling them. The Medical Research Council's Social Psychiatry Unit in the United Kingdom commissioned the development of the technology.

Specialized areas with well-defined problem-solving methodologies have led to the majority of ESs being created for such areas. MESs now suffers from the following flaws:

a. There are times when the necessary information is not accessible because specialists cannot properly explain their expertise. As a result, extracting and encoding the knowledge is a challenge.
b. It is impossible to program common sense (prudence and good judgment).
c. The majority of expert systems are not adaptable, and as a result, they only gain limited knowledge from their experiences. Therefore, the development of an expert system could need consistent attention.
d. It might be difficult for expert systems to determine whether a challenge is outside of their area of expertise.

e. When compared with human experts, an expert system has a more restricted range of sensory experiences.

1.5 SYSTEM ARCHITECTURE

Computer software that acts on the object the ability of a human expert is known as an expert system. It is common to utilize expert systems to help non-experts in circumstances when a human expert is missing. Three components make up an expert system:

a. As the name suggests, a user interface is what enables non-expert users to interact with the expert system and get guidance. As much as feasible, the user interface is meant to be as basic as possible.
b. Facts and regulations may be found in a "knowledge base." Human specialists provide information to the knowledge base.
c. Using an inference engine, the knowledge base is searched for data that fits the user's query.

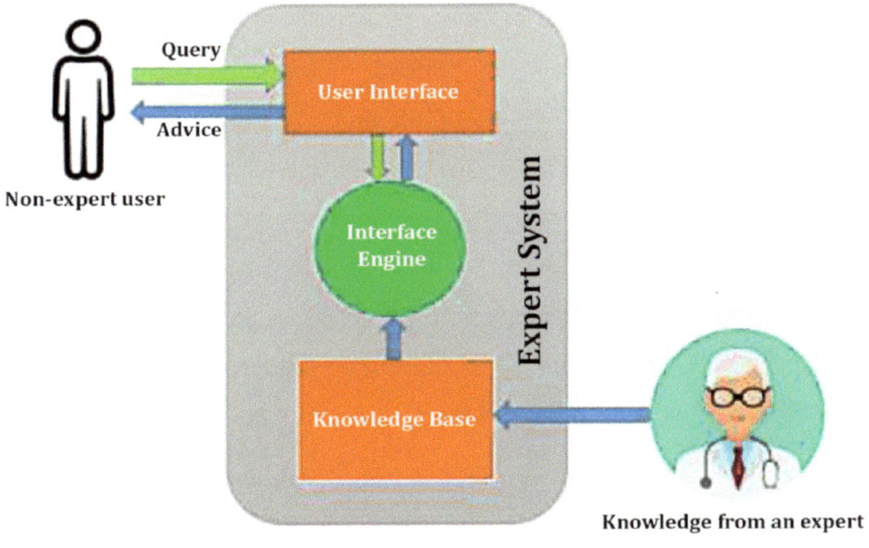

FIGURE 1.1 Expert systems for Healthcare.

When a non-expert asks a question of the expert system, A query may be asked of the expert system, and a response can be provided. When a user enters a query, the inference engine searches its knowledge base and returns an answer or advice as shown in Figure 1.1. A structure of an expert system is shown in Figure 1.2.

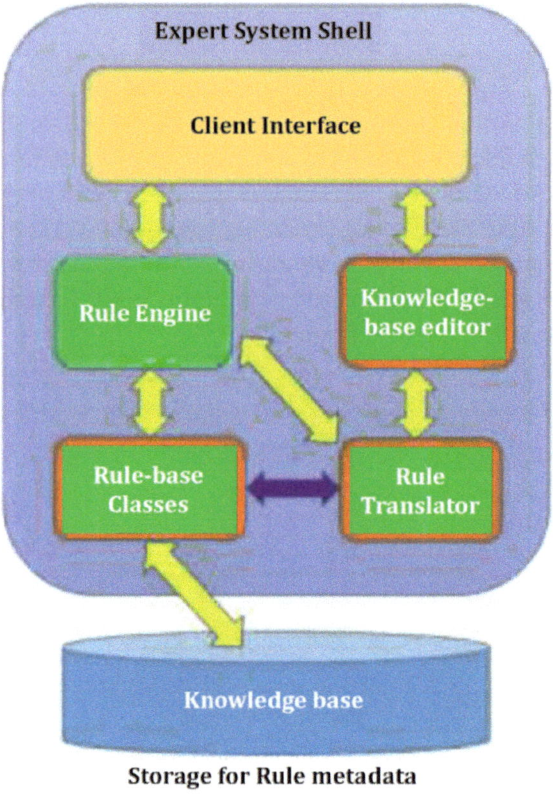

FIGURE 1.2 Structure of expert systems.

The expert system's functioning may be divided into three stages:

Acquisition of knowledge: An important part of this process is the interaction between the knowledge engineer and experts in the subject who may provide input and advice. Medics at the school's health center were contacted to get expertise and information on the most prevalent diseases that pupils suffer. Expert systems use a variety of interfaces to enter and store the collected data.

Querying or consulting with another person: The user interacts with the expert system by providing the relevant information, and the system answers using deductive reasoning. With this system, you may ask questions in English and get answers in return through a backward chaining (deductive reasoning) approach.

Results: Using this method, the expert system provides results and responses based on the queries supplied with adequate explanations of how it was able to come to these conclusions, and they are shown via the interface.

Medical diagnosis expert systems have an internal structure that consists of a knowledge base, a rule base, and an inference engine. The rule base holds the rules of inference (mostly IF-THEN rules) that are used in reasoning/decision-making; the inference engine controls how the IFTHEN rules are applied to the base; and the knowledge base provides context for the problem domain in Figure 1.2.

1.6 ETHICS AND CHALLENGES

There are several advantages associated with the use of expert systems by medical practitioners to aid in the treatment of patients. To begin, the knowledge of the systems is first retrieved in an automated fashion from the patient data. As a result, medical personnel no longer have to devote a significant amount of their time and effort to the process of building and maintaining the knowledge base. When building or maintaining the knowledge source, the sole activity that should be performed is to train the model using data. Second, the suggestion is the most effective course of action for a certain person. Thirdly, the retrieved information is more relevant to the actual world. As a result, the decision support system that has been provided in this work is intended to assist medical practitioners in providing effective clinical therapy. This is a continuing field of study, and there are various professionals working on the construction of usable expert systems as well as execution engines. This primary notion of this study is targeted at enhancing the quality of results, lowering the costs of providing healthcare, and enabling personal healthcare.

It has also been highlighted that there is a need for a decision model that tackles the factors of contextualization of guideline implementation as well as the management of incomplete and ambiguous information. However, the views of healthcare professionals have their own weight and should be included into every stage of the process of developing the knowledge base for the expert system. In addition, one of the problems with the rules at the moment is that they are too strict and do not provide room for creativity and adjustment. This decision model of the expert systems must be adaptable so that it may include recently implemented modifications.

1.7 CONCLUSION

This approach may assist the general public identify the kind of disease based on the symptoms that are presented, and because early identification can allow for the avoidance of major sickness, it can also help the public avoid being seriously ill. People who are afflicted with a particular illness may find this system to be an informative and helpful reference resource. As a result of this research, a number of benefits have been identified.

Among them are the following:

a. Because the system will diagnose the issue based on the information that it receives from the user, the outcome of this method will be free from bias and accurate.
b. The method approach makes room for human-like thinking, which leads to the production of a result that is more natural.
c. The use of an online database was efficient and successful in storing vast quantities of information and facts for inferencing purposes over the Internet.
d. Due to their rigid nature and the confines of their knowledge bases, expert systems are unable to handle exceptional circumstances or issues that are particularly difficult to solve. Because users may communicate with a human expert through tele-diagnostic equipment, this constraint may be less severe if users have access to the Internet.

In conclusion, it is anticipated that the system will be of assistance to a number of different parties, most notably members of the general public who are afflicted with a particular ailment and medical professionals who are engaged in the treatment of patients. Additionally, information on arthritis, thalassemia, and pneumococcal disease is provided through this system. The goals that were established for this system have been effectively met. During the process of developing this system, a variety of approaches are used in order to cater to the requirements of both consumers and medical professionals. Despite this, there are a few holes in the system that need to be patched up before it can function as efficiently as possible.

KEYWORDS

- **artificial intelligence**
- **knowledge-based systems**
- **machine learning**
- **data mining**
- **decision support systems**
- **Healthcare**

REFERENCES

1. Godfrey, C.; Harrison, M. B.; Lysaght, R.; Lamb, M.; Graham, I. D.; Oakley, P. Care of Self - Care by Other - Care of Other: The Meaning of Self-Care from Research, Practice, Policy and Industry Perspectives. *Int. J. Evid. Based Healthc.* **2011,** *9* (1), 3–24. https://doi.org/10.1111/j.1744-1609.2010.00196.x
2. Izadikhah, M. A Fuzzy Stochastic Slacks-Based Data Envelopment Analysis Model with Application to Healthcare Efficiency. *Healthc. Anal.* **2022,** *2,* 100038. https://doi.org/10.1016/j.health.2022.100038
3. Li, C.-C.; Shun, S.-C. Understanding Self Care Coping Styles in Patients with Chronic Heart Failure: A Systematic Review. *Eur. J. Cardiovasc. Nurs.* **2015,** *15* (1), 12–19. https://doi.org/10.1177/1474515115572046
4. Kumar, N.; Kumar, S.; Dahiya, A. K.; Kumar K. In *Application of IoT in Agriculture*, 9th International Conference on Reliability, Infocom Technologies and Optimization (Trends and Future Directions) (ICRITO), 2021, *2021,* pp 1–4.
5. Kumar, N. Kumar, K.; Dahiya, A. K. Image Restoration Using a Fuzzy-Based Median Filter and Modified Firefly Optimization Algorithm. *J. Adv. Sci. Technol.* **2020,** *29* (4s), 1471–1477.
6. Shukla, H. S.; Tripathi, R. P.; Kumar, N. Gaussian Noise Filtering Techniques using New Median Filter. *Int. J. Comput. Appl.* **2014,** *95* (12), 12–15.

7. Kumar, N.; Tiwari, A.; Dahiya, A. K.; Shukla, H. S. In *A Dual Ascent-Based Median Filter for Image Restoration*, International Conference on Advanced Computing and Software Engineering (ICACSE-2019), 2019, *2,* pp 432–435.
8. Pandey, H.; Mishra, A.; Kumar, D. Various Aspects of Sentiment Analysis: A Review. *SSRN Electron. J.* **2019**. https://doi.org/10.2139/ssrn.3349011
9. Kumar, N.; Dahiya, A. K.; Kumar, K. Modified Median Filter for Image Denoising. *Int. J. Adv. Sci. Technol.* **2020,** *29* (4s), 1495–1502.
10. Shukla, H. S.; Kumar, N.; Tripathi, R. P. Median Filter-Based Wavelet Transforms for Multilevel Noise. *Int. J. Comput. Appl.* **2014,** *107* (14).
11. Kumar, N.; Shukla, H. S.; Tripathi, R. P. Image Restoration in Noisy free Images using Fuzzy-based Median Filtering and Adaptive Particle Swarm Optimization - Richardson-Lucy Algorithm. *Int. J. Intell. Eng. Syst.* **2017,** *10* (4), 50–59.
12. Kumar, N. Role of IoT to Avoid Spreading of COVID-19. *Int. J. Intell. Networks* **2020**. https://doi.org/10.1016/j.ijin.2020.05.002
13. Shukla, H. S.; Tripathi, R. P.; Kumar, N. Image Restoration using Modified Binary Particle Swarm Optimization Richardson-Lucy (MBSO-RL) algorithm. *Int. J. Appl. Eng. Res.* **2015,** *10* (22), 43077–43081.
14. Arbaiy, N.; Sulaiman, S. E.; Hassan, N.; Afip, Z. A. In *Integrated Knowledge Based Expert System for Disease Diagnosis System*, IOP Conference Series: Materials Science and Engineering **2017,** *226,* 012097. https://doi.org/10.1088/1757-899x/226/1/012097
15. Alanazi, A. Using Machine Learning for Healthcare Challenges and Opportunities. *Inform. Med. Unlocked* **2022,** *30,* 100924. https://doi.org/10.1016/j.imu.2022.100924
16. Van Dongen, S. I.; Stoevelaar, R.; Kranenburg, L. W.; Noorlandt, H. W.; Witkamp, F. E.; van der Rijt, C. C. D.; … Rietjens, J. A. C. The Views of Healthcare Professionals on Self-Management of Patients with Advanced Cancer: An Interview Study. *Patient Educ. Couns.* **2021,** *105,* 136–144. https://doi.org/10.1016/j.pec.2021.05.021
17. Siala, H.; Wang, Y. SHIFTing Artificial Intelligence to be Responsible in Healthcare: A Systematic Review. *Soc. Sci. Med.* **2022,** *296,* 114782. https://doi.org/10.1016/j.socscimed.2022.114782
18. Bärlund, M.; Takala, L.; Tianen, L.; Kellokumpu-Lehtinen, P.-L. Real-World Evidence of Implementing eHealth Enables Fluent Symptom-based Follow-up of a Growing Number of Patients with Breast Cancer with the Same Healthcare Resources. *Clin. Breast Cancer* **2021**. https://doi.org/10.1016/j.clbc.2021.09.005
19. Ahsan, M.; Siddique, Z. Industry 4.0 in Healthcare: A Systematic Review. *Int. J. Inf. Manage. Data Insights* **2022,** *2* (1), 100079. https://doi.org/10.1016/j.jjimei.2022.100079
20. Takeshita, K.; Takao, H.; Imoto, S.; Murayama, Y. Improvement of the Japanese Healthcare Data System for the Effective Management of Patients with COVID-19: A National Survey. *Int. J. Med. Inform.* **2022,** *162,* 104752. https://doi.org/10.1016/j.ijmedinf.2022.104752
21. Yang, D.; Zhou, J.; Chen, R.; Song, Y.; Song, Z.; Zhang, X.; et al. Expert Consensus on the Metaverse in Medicine. *Clin Ehealth* **2022,** *5,* 1–9. https://doi.org/10.1016/j.ceh.2022.02.001
22. Pufahl, L.; Zerbato, F.; Weber, B.; Weber, I. BPMN in Healthcare: Challenges and Best Practices. *Inf. Syst.* **2022,** *107,* 102013. https://doi.org/10.1016/j.is.2022.102013

23. Yazdani, S.; Lerner, C.; Kulkarni, D.; Kamzan, A.; Henry, R. A New Expert System with Diagnostic Accuracy for Pediatric Upper Respiratory Conditions. *Healthc. Anal.* **2022,** *2,* 100042. https://doi.org/10.1016/j.health.2022.100042.
24. Kumar, K.; Saini, G.; Kumar, N.; Kaiser, M. S.; Kannan, R.; Shah, R. Prediction of Energy Generation Target of Hydropower Plants Using Artificial Neural Networks. In *Sustainable Developments by Artificial Intelligence and Machine Learning for Renewable Energies*; 2022, pp 309–320. https://doi.org/10.1016/b978-0-323-91228-0.00005-7
25. Kumar, K.; Kumar, A.; Kumar, N.; Mohammed, M. A.; Al-Waisy, A. S.; Jaber, M. M.; Shah, R.; Al-Andoli, M. N. Dimensions of Internet of things: Technological Taxonomy Architecture Applications and Open Challenges—A Systematic Review. *Wirel. Commun. Mob. Comput.* **2022,** *2022,* 1–23. https://doi.org/10.1155/2022/9148373
26. Kumar, K.; Singh, R. P.; Ranjan, P.; Kumar, N. Daily Plant Load Analysis of a Hydropower Plant using Machine Learning. In *Algorithms for Intelligent Systems*; 2021, pp 819–826. https://doi.org/10.1007/978-981-33-4604-8_65
27. Sharma, A.; Kumar, N. Encryption of Text using Fingerprints as Input to Various Algorithms. *Int. J. Sci. Res.* **2014,** *3* (4), 418–421. https://www.ijsr.net/archive/v3i4/MDIwMTMxNDYx.pdf
28. Kumar, K.; Shah, R.; Kumar, N.; Singh, R.P. Application of Robotic Process Automation. In *Applications of Artificial Intelligence in Engineering. Algorithms for Intelligent Systems*; Gao, X. Z., Kumar, R., Srivastava, S., Soni, B. P., Eds.; Springer: Singapore, 2021. https://doi.org/10.1007/978-981-33-4604-8_75
29. Kumar, N.; Shukla, H. S.; Tiwari, A. K.; Dahiya, A. K. Dual Ascent based Median Filter for Image Restoration. *SSRN Electron. J.* **2019**. https://doi.org/10.2139/ssrn.3350334.

CHAPTER 2

DECENTRALIZED MANAGEMENT OF HEALTHCARE IoT DEVICES FOR TRANSFORMING THE FUTURE

MINAKSHI SEHRAWAT and DEERGHA SHARMA

The NorthCap University, Gurugram, Hariyana, India

ABSTRACT

Healthcare professionals know the worth and relevance of adaptability and resilience in managing their clinical operations during their day-to-day patient care. It is urgent for enhanced care to- document, record, and track the vital patient data and information for clinical staff to directly navigate gauge and make essential compliance. When traditional ways are slow, disruptive technologies have the potential to add value to the system. It can comprehend health professionals' expertise by bringing all features on one platform, starting from record-making, doctor's prescriptions, diagnosis recommendations, critical treatment alerts, OT occupancy status, reporting, and collaboration. Medical units can customize and configure the agile software as it suits them the simplest. Decentralized management system of Healthcare IoT devices is one such disruptive practice widely sought after in Healthcare IT field.

Well, now the question is how far a Healthcare expert shall go to decentralize the process, functioning and even decision-making? The question shall be answered by telling the benefit and every aspect of the decentralizing in closer association with the improved patient care. Advanced OT Scheduling solutions, Queue Management System, Digital Signage System,

IoT-Enabled Healthcare Systems: Applications, Benefits, and Challenges
Yatindra Kumar, Shashi Kant Verma, Archana Singh, Krishna Kumar, & Manoj Gupta (Eds.)

Asset Management System, OT Control Panel, etc. can easily integrate with clinical, financial, and regulatory tools to bring the real data on the table and make the job of operating patients smoother and easier. Through this type of smart devices and solutions, patients can continuously track their health condition, monitor changes, receiving alerts, and consult with the doctor from their comfort zone. The digital revolution and abrupt transformation have made this journey seamless through video conferencing, phone–video connections, continuing remote medical information and education, patient portals, nursing call centers, patient-focused applications, and much more. Thanks to information and technology advancement, which has addressed the concerns of many patients who were earlier devoid of consultation due to their inability to travel or frequently travel to hospitals or clinics to concerning health issues.

This chapter explained the pitfalls of the Healthcare industry to solve them while connecting quickly with customers, less upfront investment, and delivering promises even remotely. Further, how resilient intuitive solutions will help making paths clear for traditional and contemporary business strategies by taking charge of consolidating your data, streamlining business operations, designing workflows, decreasing documentation, and handling perpetual data confidently. It is a high time when most healthcare providers are planning, implementing, and democratizing the disruptive technology to propel their businesses and take the entire industry forward.

2.1 INTRODUCTION

Disruptive technology is rapidly reshaping the workflow practices in healthcare institutions. Especially when COVID-19 entered and penetrated widely across the world, online work practices and techno-led services helped physicians pool their resources and offer the best alternatives. Decentralized work practices including advanced AI-enabled applications and advance computing solutions are woven to make impactful use of technology for growth. For instance—advanced queue management solutions for taking online appointments, tracking status in the queue, and accessing real-time queue information without actually standing in it. The hospital's marketing team can further reach out to patients for important announcements, alerts, and advertisements and add revenue through targeted marketing using technology-based support. It is able to map patients' journey, manages wait time, and organizes their flow in a hassle-free and comfortable environment. Meanwhile, you can also obtain direct feedback on how well your patients have been treated in your hospital and get appreciated immediately. Alongside a variety of preferences, clinics and clinical associations can use the troublesome innovation that empowers simplicity, solace, and accommodation. Further, the client will feel more certain and trusted with your well-being image.

2.1.1 RELEVANCE OF DECENTRALIZED PRACTICES

The current worldwide pandemic situation has increased the importance of decentralized solutions to the operations of medical practices. Integrated practice management tools that carry out crucial administrative and billing tasks, but their main attention is on the clinical aspect of daily operations. It is a high time when most healthcare providers are planning, implementing, and democratizing the disruptive technology to propel their businesses and take the entire industry forward. While other operators are battling with capacity crunch, IoT-based enablers have enough space, flexibility, and mechanisms to co-create business, fill the gap, and transform the fundamental capacities with core technologies. Decentralized healthcare solutions have built-in intelligence that makes it possible to automate a number of medical procedures.

2.1.2 CONCEPTUAL FRAMEWORK

Internet of Things (IoT) is one of the technological advancements in computing and communication comprising web-enabled smart devices containing physical components, such as actuators and sensors and processors to collect, process and send data perceived from external physical environment.[1] By connecting with IoT gateways or other edge devices, the sensor data are shared by IoT devices. At times, these devices communicate with other associated devices and process the information received.

A published report by Statista has forecasted that the usage of IoT-enabled devices would increase 75 billion by 2025. The IoT dimensions opened up new paths and opportunities in various industries by fusing the benefits of edge computing, fog computing, communications technology advancements, cloud computing, and sensors.

IoT devices are gradually adopted in Healthcare and wellness services. IoT has a significant contribution in reshaping Healthcare into smart Healthcare. Following are the prerequisites of smart Healthcare.

1. Treatment of a specific person at appropriate time.
2. Data availability at readily available place.
3. Accurate treatment of illness.
4. Accessibility of Healthcare services at remote locations.
5. Improved efficiency by reducing operational cost.

The Healthcare sector functions through five segments: Pharmaceuticals, Diagnostics, Medical equipment and supplies, Telemedicine, Hospitals. IoT has found its applications in telemedicine due to development of recent technologies in internet and telecommunication. Telemedicine has bridged the gap between urban and rural Healthcare by providing remote diagnostic facility and lower consultation cost and potential of tapping remotest of area. IoT has also transformed home Healthcare by providing high-quality healthcare services at lower cost. The advancements of mobile technologies for instance 4G and 5G applications of mobile computing are advancing in Healthcare services. Due to the significant influence of IoT on the Healthcare business and its applications, the main contribution of this study is to review the following features of IoT in Healthcare.

IoT technology is used in many different applications and is expanding remarkably in several industries. Yet, it lacks a specified standard architecture that might be used anywhere. IoT's functionality and processes in many industries are key components of its architecture. Yet, Figure 2.1 shows the IoT's fundamental architecture. The architecture is divided into four layers: the application layer, the data processing layer, the network layer, and the sensing layer. The following is a discussion of the layers:

1. **Sensing Layer**

   The sensing layer comprises of sensors, actuators and devices. Identification technologies and perception are the key elements of IoT. The sensors and actuators accepts data on environmental and physical parameters, process and emit that data over the network.

2. **Network Layer**

   The layer incorporates Data Acquisition System (DAS), Network/ Internet gateways. The DAS assembles the analog data of sensors and convert it into digital data.

3. **Data processing Layer**

   It is a processing unit of IoT ecosystem wherein data is analyzed and pre-processed before remitting it to data center.

4. **Application Layer**

   Data is maintained and used by a variety of end-user applications.

2.2 IoT-BASED HEALTHCARE ARCHITECTURE

The IoT architecture in Healthcare are divided in three layers. The layers are their associated health implications are discussed as follows:

2.2.1 PERCEPTION LAYER: SENSING SYSTEMS FOR DATA COLLECTION

Perception and identification technologies are the key elements of IoT. The perception layer comprises sensors that perceive variations in the surroundings and include, for instance, Infrared sensors, GPS, cameras, radio frequency identification (RFID), and smart devices sensors. These sensors permits extensive perception utilizing object recognition, geographical recognition, and location recognition and convert this information to digital signals, which is more feasible for network transmission.[11] The sensor technologies permit real-time monitoring of the treatments and facilitates the acquisition of multiple psychological parameters regarding a patient to fast-track diagnosis and high-quality treatment. Bluetooth-enabled inhalers, digital medications, smart voice assistants are some of the life-saving IoT sensor devices.[12]

2.2.2 NETWORK LAYER

For the healthcare application of IoT, network layer technologies contain wired and wireless networks for communicating and storing processes information either at a centralized location or locally.[11] The evolving 5G technology has the potential to connect number of devices.[13]

The delivery of health services by cloud-based computing is advantageous due to the flexible and scalable data acquisition, storage, and transmission.[14] The cloud facilitates medical IoT devices, such as smart phone apps, patient portals, and big data analytics-driven decision support system.[15]

Edge cloud is one of the evolving cloud computing approaches that permits IoT sensors and network gateways for processing and analyzing the data at the edge in a decentralized manner reducing the traffic of data transmission.[16]

2.2.3 APPLICATION LAYER

The application layer provides application-based services to the users. IoT-driven healthcare services and usage of deep machine learning facilitate medical professionals with enhanced diagnosis capabilities. Although, diagnostic confidence shall never reach 100%, blending clinical expertise and machines reliably enhances system performance. Deep learning and AI can optimize disease management, provide analysis produced from IoT devices health apps and witnessing adoption in Healthcare.[17] Some examples include risk prediction, mental health,[18] future medical outcomes, and anticipate the progression of bone diseases[19] and cardiac arrhythmias.[20] The popular use cases are shown in Figure 2.1.

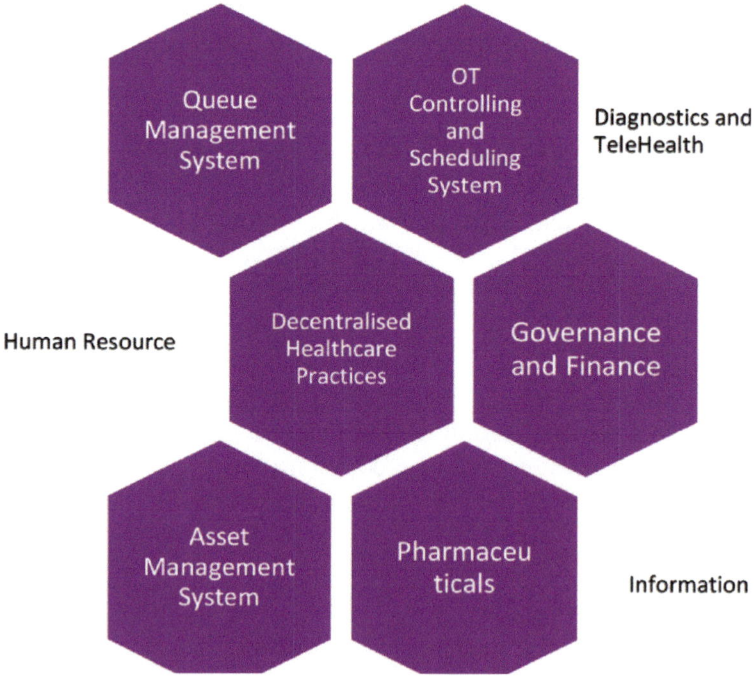

FIGURE 2.1 Applications of IoT in Healthcare.

Here is the comprehensive suite of professional services:

a. Queue Management System

Distantly scheduling patient: A specialist can exploit programming by re-appropriating the tedious undertaking of patient planning. Plus, scheduling is a basic job in customer experience. It permits consistent activities even with satisfactory offices to oblige rescheduling, repeating arrangements, or retractions. If the current innovation does not fuse, say—the abrogation, it can hurt the income management and create a blunder feeling. More powerful planning arrangements can assist the specialist with investigating designs like patient show designs, scratch-offs focuses, and the days and seasons of most extreme arrangements. It will additionally reinforce the information-based dynamic of the specialists.

Correspondence with visitors: The functionaries of QMS programming empower smooth correspondence between the medical services groups and patients. Utilizing the solution, patients can sign in from any web-associated gadget, see the forthcoming timetable, read past clinical history, send messages with incorporated protection highlights, or view their status calmly. Likewise, this system gives the freedom of tasks like rounding out their web-based enrolment or clinical history structures, saving clinical records, inpatient experiences, etc.

Manage multiple operations simultaneously: QMS solutions get hassle-free work choices. It integrates with your core business and helps your customers manage operations like—signing in, notifying, ticket generation or transfer, position tracking, and telling them estimated wait time. Further, they can leave their feedback and suggestions based on their service experience. Meanwhile, it helps you generate business insights with the help of your daily-in and out data. Through this cost-effective, comprehensive QMS multitude business support, you can generate powerful analytics and reports. Such dashboards and statistics will help you make improved decision-making and generate valuable insights to grow more.

Create and Send Regular Notifications: Queue Management System will help you please your customers a little more and grab their loyalty. Your hospital staff can also use this facility for tailoring the specific workflows and integrate them through SMS or other digital modes. Inbuilt features of QMS make this use easier via the latest technology-driven, web-based, or handheld mobile devices like phones, tablets, laptops, or desktops using cloud-based software. Also, it eliminates chances of human-error during appointment. As, one of the significant patients dealing way in hospitals is an appointment but it has human factors involved and do not offer resilience to unexpected and chaotic issues. Conversely, all-inclusive Cloud QMS software can facilitate the queuing process and flow in a customized way. It helps their seamless movement without needing to stand them physically and wait in line.

b. Electronic Medical Record (EMR) Solutions

Many medical tasks can be automated because to the intuitive intelligence that is embedded into electronic medical record (EMR) solutions. These solutions mark a dramatic shift away from the paper-based or analog medical industry. EMR software can manage a smooth workflow, track data promptly, make it available, and produce crucial records that will aid numerous healthcare practitioners in providing their patients with the greatest guidance possible. Moreover, EMR systems improve results while lowering operating costs and encouraging increased patient interaction. The advantages of using EMR solutions is, it can retrieve patient data immediately, enhance patient participation, improved patient Healthcare, streamline your medical practice, reduce medical errors, save costs, and improve efficacy.

EMR software compiles all patient data in one location. Examples include the medications they take, the tests they've had done in the past, the recommendations of their doctors, and the medical practices they adhere to. It enables prompt retrieval of any patient's necessary information and aids the doctor in lowering the danger of negligence in any fictitious situation.

In modern settings, the vast majority of patients are technologically savvy. Patients today will appreciate the change from physical to digital versions of their medical records. Additionally, it has a self-service element and gives consumers the ability to control their health conditions. Users can book appointments, view clinical data from their medical tests, obtain essential information, and receive follow-up from medical personnel.

Healthcare consumers and providers can communicate and interact more effectively thanks to EMR technologies. EMR provides specialized solutions to meet each client's particular demands. Before making a purchase, you must be clear about your demands and priorities so that the contractor is aware of your top priorities for features and promises. Medical professionals will be able to work in a more structured manner thanks to EMR systems, which will give them access to complete and correct data.

Improved information accessibility will lead to the appropriate understanding, reduce the likelihood of ambiguity, and enable practitioners to understand the linked and unrelated elements of health disorders. It decreases misunderstanding gaps, mistakes, and irrational prescriptions or test suggestions. Improved management and results will result from enhanced information and activity synchronization across many departments. The ability to accept input, combine data, and produce analysis enables medical units to reduce costs and time commitments while also increasing efficacy. For instance, the account department and the patients can review the history of consultations, medications, and procedures when an invoice is generated at the hospital's billing counter. Also, it eliminates the possibility of data duplication, upholds transparency, and results in mutual pleasure.

Even if a patient is not in need of any medical attention, an EMR system still functions. The most anyone can do is to have their information in one place and be able to quickly access it when needed because there is no way to prepare for any medical emergency or event.

2.3 ASSET MANAGEMENT SYSTEM

AMS is an entirely blended tool to incorporate asset arranging and management with business strategies and budgetary cycle in medical hospitals, health centers, or hospital care establishments. It tracks, screen, oversee, and handle assorted assets to help you make administrative consistence, decrease operational expense, and convey the most excellent patient consideration. Healthcare institutions have volumes of things and assets that should be overseen, evaluated, and refreshed intermittently. The mounting pressing factor of patient consideration, administrative consistency, and monetary strains has made the asset management framework a vital segment of their prosperity.

It encourages a reasonable and organized path for dealing with the hospital's assets in the long haul for their valuation, buy, activity, and furthermore up to their deal. This fundamental arrangement decreases your stresses over dealing with various assets and gear accessible in the hospital, following their area/status, contracted or surplus things, and considerably more. In nutshell, it allows

a. Identification of the necessity of assets
b. Take record of the existing pattern of assets, advantages, and danger included
c. Operation and support cost-related
d. Duration and cost of removal of the asset
e. Know asset accessibility and its use
f. Capitalize on asset's undiscovered potential

Advantages of picking the correct Asset Management System can be huge. Plan and oversee assets in the hospital continuously. Innovation keeps the paperless record of things and alarms the doctors about their excess, deficiency, or even area at whatever point they need them. During understanding associations in pivotal cases, a specialist cannot bear to sit around idly looking through the necessary pieces physically; all things are considered, they will require them progressively. Therefore, the assets should be cleaned and prepared to utilize. AMS encourages telling the "ability" of each object to experts continuously. Accepting a without paper component encourages you to influence innovation while supporting the green cause and keep up cleanliness through sans touch systems. Our product gathers all data in one spot. For example, things purchasing subtleties, solicitations, in-stock or out-stock things, things that need overhauling/fixing/cleaning/supplanting, and so forth, lapsed, area of any hardware, disposed of pieces, assets arranging, and so on. It permits recovering any necessary data promptly and helping the doctor decline operational expense, time, and danger generally. AMS programming can deal with the impeccable work process, track information on schedule, make it open, and produce imperative asset records to control different hospital care experts in contributing the best administrations to their patients. Utilizing the AMS arrangement, keep your data set precise, refreshed, and updated in light of the fact that nobody can plan for any health-related crisis. The best is to keep all data in a solitary place and recover them quickly as and when required. The possibility to follow input/yields, coordinate, and recover information will encourage hospital units in expense and efficiency and effectiveness improvements. Further, it disposes of the odds of information duplication, guarantees straightforwardness, and brings fulfilment at all levels. This hospital asset progress from an actual structure to a computerized state is an inviting move for expanding income. Ideal accessibility of things or their data helps acquire the correct knowledge, decline odds of uncertainty, permit experts to limit the arrangement hole and lessen blunders and pointless worries. Because of improved data management, the exercises between various units will encounter better coordination and results. Then, it is empowered the expiry management framework to help forestall terminated things and supplant them with a new one.

2.4 TELEHEALTH SOLUTIONS

With this approach, technology is quickly taking over the responsibilities of medical specialists and nursing staff. It enables professionals to focus more on the patient's health and spend less time synchronizing paperwork and documentation. The operational, clinical, and billing workflows are streamlined by the intuitive features that come standard with telehealth software, helping to sustain a healthy demand for Healthcare. The ability of the newest technology-enabled solutions to efficiently handle the repetitive nature of administrative work makes a significant impact at hospitals. Your patients will have a painless medical experience, which will inevitably draw in more patients and make using your services more convenient.

The use of telehealth solutions for senior life has many benefits. Clinics and healthcare facilities can benefit from the innovative technology that makes life easier, more comfortable, more convenient. The user will also feel more secure and trusted when using the telehealth services provided by net health. Software enables therapists to outsource the time-consuming task of patient scheduling. Moreover, scheduling is crucial to the client experience. It enables smooth operations even when there are sufficient facilities to handle cancellations, recurrent appointments, or rescheduling. If the current system does not support, let us say, cancellation, it could affect cash flow management and create an error-prone environment. The therapist can evaluate trends, such as patient show patterns, cancellations points, and the days and times of the most appointments with the use of more powerful scheduling solutions. It will support the therapists' data-driven decision-making even more.

Smooth communication between healthcare teams and patients is made possible by the features of telehealth software. Patients can use the software to log in from any internet-connected device, access their payment information, read their past medical history, send messages with built-in privacy safeguards, and preview the forthcoming schedule. This paperless system gives users the freedom to complete online enrolment or medical history forms, save medical records, manage inpatient accounts, and other tasks. Patients will benefit from digitization in a secure manner. Medical patient records can be created, uploaded, edited, saved, or deleted using telehealth for senior living. The ability to take digital notes is one of its features that enables patients or doctors to take notes as needed. Naturally, it encourages productivity while removing the possibility of losing or harming paper. These technologies make it easier for experts to handle the patient's extremely sensitive or confidential information and prevent its disclosure. Elderly living can now choose from hassle-free jobs in telehealth. The tools needed to collect the money owed by doctors are all included in these systems. It includes the ability to easily construct and deliver statements and invoices to patients. Additionally, it entails making claims to insurance companies simpler and monitoring the progress of the claim payment. A disruptive mechanism like this one frees up the patients' time and energy so they may concentrate on their main goal. Additionally, features like payer configuration for simple billing, optimal reimbursement, and appeal management aid in streamlining payments in the proper manner.

The program offers check-in procedures that make patients feel comfortable while outlining every session's preparation in depth to enhance their experience with your clinical services. On the one hand, it increases staff production while maximizing their efficiency. Better outcomes are also a result of having access to the best clinical content library, flexible mobility, offline access, etc. In order to regain the trust of the senior living service users, it enables the procedure to be simplified while maintaining transparency.

The centralized online telehealth solutions for senior living further become authentic by its characteristic under which physicians can sign from anywhere, anytime. Technology gives wings to it. With this, they enjoy mobility and avoid duplication of work by getting centralized access to all documents in one place. It saves the time and efforts of physicians while allowing them on their core area of patient care. Moreover, it satisfies the clinicians and streamlines the workflow with much satisfaction. Telehealth software allows the users to make availability of all this information which are crucial for local users through offline access and best clinical library database, to drive efficiency and better outcomes.

2.5 OT SCHEDULING SYSTEM

Operation Theater Scheduling System helps in setting and managing the schedule of operations/surgeries for patients. It keeps a record of all the activities and engagements in OT and checks for its next availability. Depending on the different medical units' additional requirements, you can customize the features of OT scheduling systems. Largely, it stores all OT information at centralized place, retrieves them repeatedly, and schedules and blocks OT as per the situation. It contains complete information of surgery, such as date, time, type, hours of operations, blood group, name of the surgery, swabs count, complications detail, amount of fluid given, conditions of exigencies, etc. Saves time, effort, and energy: Having kept all data and the information in place, the OT scheduling software system will provide you the best coordination- before, during, and after- your surgical time. Its suitable information storage mechanism will keep the record of each detail, which you may or even may not require. No doubt, this is going to ease your job, save your energy, and let you worry merely for your patient's health.

There can be varied nature of situations. For instance, say—with Two OTs, Three Surgeons, Four Staff working in different shifts, One Anesthesia Provider coming for different intervals, etc.; more than five operations are to be scheduled in the next 3 days. In such a situation, like all other healthcare professionals, you will feel frustrated while checking for each patient's best solution and compliance. You may have to sit for another hour to talk to each one and fix the issue manually. Nevertheless, the software mechanism is uniquely made to address this challenge and let you set free for your core business. OT Scheduling systems will cross-check with your free time slots, unoccupied personnel, and vacant space along with duration to ensure that you commit to your patient the best timing for their forthcoming treatment.

Allows to take all the details on records: There can be a number of pleasant or unpleasant experiences with the surgery. Furthermore, there can be situations like an extra demand for any form of inputs, maybe any staff is arriving late or absent due to some issue! In such cases, you might require the immediate help of knowing what next could be possible. During such situations, OT Scheduling System has the potential to rescue you and take care of your extraordinary demand.

2.6 OPERATION THEATER CONTROL PANEL

OT Control Panels are innovative methods of controlling the entire OT settings just with a simple touch screen. In these intuitive systems, controls of different services required in OT are incorporated into a single interface panel, including lightning, ventilation, temperature and humidity, and critical alarms. More straightforward to use and sleek in design, these modular OT systems are helpful to keep pace with technology and the latest equipment in the medical industry. Made with superior quality raw material and advanced manufacturing techniques, these systems are directly mounted on the theater wall that can be controlled with merely one touch of a fingertip. If you are looking for affordable, user-friendly, and high-quality Operation Theater Control Panels, Pal Software will facilitate you well.

The foremost priority of any healthcare institution is to offer the best care to their patients. Each effort of a physician keeps the "patient's health" central. Moreover, that is possible when no detail is missed, and no control system is loosely handled. Bespoke solutions like Advance OT Control Panels are integrated systems that make operation theater processes easier for surgeons while minimizing the chances of error and maintaining the hygiene level at its best. Integrate all control systems in one interface: It will combine different accessories in a single panel and thus reduce the efforts to maintain each of them separately. Meanwhile, it will also reduce the space required and the cost of their installation and management. Systems that can be built on-demand, and customized way: Our Solution development team builds, configure, and integrate customized OT control systems as desired by our clients.

2.7 CONCLUSION

Technology has many uses and has diverse connotations for different people. Technology has drastically changed the world we live in and improves its efficiency. Having total and exact information close by will permit the medical professionals to work in a more coordinated manner. Additionally, the work delays because of non-prepared, expiry, or lost status of the gear are diminished. Utilizing this arrangement and its key functionalities, you can completely remain serious and receptive to your empowering business drivers.

This disruptive mechanism saves a lot of time and energy for the patients and focuses on their core tasks. Moreover, it improves the capacity of clinicians and patients by saving time, cash, and energy. Clinical units are progressively disposed to depend on such decisions, which will give them a return in the more drawn outrun. Additionally, the highlights of queue management practices improve your clinical businesses on all the mainstream stages. Survey observing will create an investigation and assemble believability. Clinicians will discover them in an upbeat state with the correct review generation and consistency.

Source: Naresh et al.[2]

KEYWORDS

- **Healthcare**
- **IoT**
- **asset management**
- **hospitals**

REFERENCES

1. Domingo, M. C. An Overview of the Internet of Things for People with Disabilities. *J. Netw. Comput. Appl.* **2012,** *35* (2), 584–596.
2. Naresh, V. S.; Pericherla, S. S.; Murty, P. S. R.; Sivaranjani, R. Internet of Things in Healthcare: Architecture, Applications, Challenges, and Solutions. *Comput. Syst. Sci. Eng.* **2020,** *35* (6), 411–421.
3. Wu, M.; Lu, T. J.; Ling, F. Y.; Sun, J.; Du, H. Y. In *Research on the architecture of Internet of Things*, 2010 3rd International Conference on Advanced Computer Theory and Engineering (ICACTE); IEEE, 2010; Vol. 5, pp V5–484.

4. Merchant, R. K.; Inamdar, R.; Quade, R. C. Effectiveness of Population Health Management Using the Propeller Health Asthma Platform: A Randomized Clinical Trial. *J. Allergy Clin. Immunol. Pract.* **2016,** *4* (3), 455–463.
5. Kim, S. H.; Chung, K. Emergency Situation Monitoring Service Using Context Motion Tracking of Chronic Disease Patients. *Cluster Comput.* **2015,** *18* (2), 747–759.
6. Dehury, C. K.; Sahoo, P. K. Design and Implementation of a Novel Service Management Framework for IoT Devices in Cloud. *J. Syst. Softw.* **2016,** *119,* 149–161.
7. Kumar, P. M.; Gandhi, U. D. A Novel Three-Tier Internet of Things Architecture with Machine Learning Algorithm for Early Detection of Heart Diseases. *Comput. Electric. Eng.* **2018,** *65,* 222–235.
8. Sebbak, F.; Benhammadi, F. Majority-Consensus Fusion Approach for Elderly IoT-Based Healthcare Applications. *Ann. Telecommun.* **2017,** *72* (3–4), 157–171.
9. Hussain, A.; Wenbi, R.; da Silva, A. L.; Nadher, M.; Mudhish, M. Health and Emergency-Care Platform for the Elderly and Disabled People in the Smart City. *J. Syst. Softw.* **2015,** *110,* 253–263.
10. Wickramasinghe, A.; Torres, R. L. S.; Ranasinghe, D. C. Recognition of Falls Using Dense Sensing in an Ambient Assisted Living Environment. *Pervasive Mobile Comput.* **2017,** *34,* 14–24.
11. Sethi, P.; Sarangi, S. R. Internet of Things: Architectures, Protocols, and Applications. *J. Electric. Comput. Eng.* **2017,** *2017.*
12. Kelly, J. T.; Campbell, K. L.; Gong, E.; Scuffham, P. The Internet of Things: Impact and Implications for Healthcare Delivery. *J. Med. Internet Res.* **2020,** *22* (11), e20135.
13. Li, S.; Da Xu, L.; Zhao, S. 5G Internet of Things: A Survey. *J. Ind. Inform. Integrat.* **2018,** *10,* 1–9.
14. Darwish, A.; Hassanien, A. E.; Elhoseny, M.; Sangaiah, A. K.; Muhammad, K. The Impact of the Hybrid Platform of Internet of Things and Cloud Computing on Healthcare Systems: Opportunities, Challenges, and Open Problems. *J. Ambient Intell. Humanized Comput.* **2019,** *10* (10), 4151–4166.
15. Dang, L. M.; Piran, M.; Han, D.; Min, K.; Moon, H. A Survey on Internet of Things and Cloud Computing for Healthcare. *Electronics* **2019,** *8* (7), 768.
16. Pan, J.; McElhannon, J. Future Edge Cloud and Edge Computing for Internet of Things Applications. *IEEE Internet Things J.* **2017,** *5* (1), 439–449.
17. Tobore, I.; Li, J.; Yuhang, L.; Al-Handarish, Y.; Kandwal, A.; Nie, Z.; Wang, L. Deep Learning Intervention for Healthcare Challenges: Some Biomedical Domain Considerations. *JMIR Mhealth Uhealth* **2019,** *7* (8), e11966.
18. Pham, T.; Tran, T.; Phung, D.; Venkatesh, S. Predicting Healthcare Trajectories from Medical Records: A Deep Learning Approach. *J. Biomed. Inform.* **2017,** *69,* 218–229.
19. Li, H.; Li, X.; Ramanathan, M.; Zhang, A. Identifying Informative Risk Factors and Predicting Bone Disease Progression via Deep Belief Networks. *Methods* **2014,** *69* (3), 257–265.
20. Mathews, S. M.; Kambhamettu, C.; Barner, K. E. A Novel Application of Deep Learning for Single-Lead ECG Classification. *Comput. Biol. Med.* **2018,** *99,* 53–62.
21. Naresh, V. S.; Pericherla, S. S.; Murty, P. S. R.; Sivaranjani, R. Internet of Things in Healthcare: Architecture, Applications, Challenges, and Solutions. *Comput. Syst. Sci. Eng.* **2020,** *35* (6), 411–421.

22. IoT: Number of Connected Devices Worldwide 2012–2025. Statista, Statista. [Online]. https://www.statista.com/ statistics/471264/iot-number-of-connected-devices-worldwide/ (accessed Apr 4, 2019).
23. 2018 Global Healthcare Outlook. [Online]. https://www2.deloitte.com/content/dam/Deloitte/global/Documents/Life -Sciences-Healthcare/gx-lshc-hc-outlook-2018.pdf (accessed: Apr 4, 2019).
24. IBEF Presentation. [Online]. https://www.ibef.org/download/Healthcare-February-2018.pdf (accessed Apr 8, 2019).
25. Jagadeeswari, V.; Subramaniyaswamy, V.; Logesh, R.; Vijayakumar, V. A Study on Medical Internet of Things and Big Data in Personalized Healthcare System. *Health Inform. Sci. Syst.* **2018,** *6* (1), 1–20.
26. Bhuvaneswari, G.; Manikandan, G. A Novel Machine Learning Framework for Diagnosing the Type 2 Diabetics Using Temporal Fuzzy Ant Miner Decision Tree Classifier with Temporal Weighted Genetic Algorithm. *Computing* **2018,** *100* (8), 759–772.

CHAPTER 3

SOLDIER STRAP: AN IoT-BASED SAFETY AND SECURITY BAND FOR SOLDIERS

SHUBHAM VERMA[1], NIKHIL SHARMA[2], ILA KAUSHIK[3], DEEPAK KUMAR VERMA[4], and MANOJ GUPTA[5]

[1]*Tata Consultancy Services, India*

[2]*Delhi Technological University, Delhi, India*

[3]*Krishna Institute of Engineering and Technology, Ghaziabad, India*

[4]*Greater Noida Institute of Technology, Greater Noida, India*

[5]*Department of Electrical Engineering, Faculty of Engineering, Chulalongkorn University, Bangkok, Thailand*

ABSTRACT

In today's world, Internet of things (IoT) has developed advanced technology and is playing a major role in society and for other developments. IoT is evolving through the past few years. In some past few years, as we have seen that IoT is used as essential source of infrastructure for various types of organizations and has developed various instruments and technology to counter the modern technique. For instance- The Advance Encryption Standard (AES-256) which is the safest technique in the present scenario for the transmission of the data. In recent years, many countries have adopted the Advance Encryption Standard (AES-128, AES-192, and AES-256) techniques for the safety and security of their country. The Advance Encryption Standard (AES-128, AES-192, and AES-256) is the

IoT-Enabled Healthcare Systems: Applications, Benefits, and Challenges
Yatindra Kumar, Shashi Kant Verma, Archana Singh, Krishna Kumar, & Manoj Gupta (Eds.)

best way to protect data because in this technique, it is very difficult to crack the encoded key and the decoded key. In this paper, IoT and Advance Encryption Standard (AES-256) is used to develop a soldier strap which is a band used for the safety and security of the soldiers. The security of our nation depends upon our soldiers and so their security and good health must be our utmost priority. A soldier strap is a device that is capable of tracking a soldier's movement, heartbeats, body temperature, and even sending surveillance images to the military base. The highlighted key feature of the device is SOS signals which can be sent by the soldier in case of an injury or ambushed by the enemies. The device works on AES data encryption technology which sends the data in an encrypted form in a symmetric block cipher. This prevents the hacking of data by outer sources. This would also facilitate communication between the soldiers. The battery life of the device would be long enough to support soldiers trapped in adverse conditions.

3.1 INTRODUCTION

As soldier goes on the patrolling over the border areas, they face difficulties to communicate with the BSS (base station). They also face many problems at that time and cannot communicate directly to the BSS for any direct help or to pass any of the information.[1] Therefore, soldier strap is one of the initiatives to resolve the problem so that they may communicate to the BSS as well as to their colleagues whenever required. This soldier strap will also be helpful to pass the message for medical help or any unusual activity noticed on the border areas. The soldier strap is capable of transmitting the data through a Wi-Fi module to the server, and BSS can directly access the message. This soldier strap has a sound module, body temperature sensor, heart rate sensor, and GPS module. By these gadgets, it will help for the BSS to search their soldiers who are on the patrolling. By the GPS locations, they may be tracked and BSS can send help to them at any cost.[2] This soldier strap also has a safety feature that by chance if the soldier is stuck in the enemy area and the enemy wants to access this soldier strap, it will not work due to two reasons, the first reason is if they will open it forcefully, then the device will automatically go in the self-destroy mode and nothing will be there in the band, whereas if they try to take the component out and try to add in their system made by them, the soldier strap will also not work because in this, the data transmission is done through Advance Encryption Standard (AES-256). Therefore, every device is when added to the soldier strap, then a unique ID is generated and it will work only with that ID; if the hardware or anything's is changed, then we have to first make contact with the server, and then only it will be operatable. So from the security point of view, we have tried to implement it as safely as possible. As in today's world, AES-256 is one of the safest ways to transmit data because to decode the data, it will take a more time but even then it will not be decoded. Therefore, for the safety and security of the nation, it will also be beneficial for Indian soldier's life as well. In soldier's security, the biosensor system provides various kinds of sensors such as small physiological sensors which are used to detect the body posture and also the physical presence of the soldier. Barometric sensor is used to measure the temperature and energy efficiency of the area, and oxygen analyzer is used to measure the concentration of oxygen level of the soldier because, at higher altitudes, the soldiers may experience oxygen shortages and a range of other problems, processing capabilities of all the devices and working duration should be efficient, and furthermore, the transmission modules should be perfect to operate in all areas in all situations while taking national security concerns in to consideration. It can further enable affordable wearable technology and also an unremarkable solutions for our soldiers' health-monitoring systems and for their improved transmission process.[3] We propose to utilize the GPS system to track every movement of the soldier and also to alert them about any problems in the particular location if they receive any information about the soldiers' direction of movement in the form of defense direction on maps that the soldiers use to locate themselves. By employing the latitude and longitude data they communicate the information to the base station and their associates, making it simple to determine the direction to easily locate them. Here, we are using AES data encryption technology that has become the global standard of encryption, and it can be implemented through software and hardware globally to encrypt a particular sensitive data as it is also an essential part for all the electronic data protection. The main objective of this project is to track the location of the soldiers and to measure the health status of the soldiers using some of the more important concerning metrics and to measure the heartbeats rate and temperature of the body. It would be also helping in determining the exact location by using one of the advanced sensors like GPS tracking system and enabled it using a GSM modem. It can also transmit all information of their areas to the base station so that some important decisions may be taken to tackle and necessary action may be taken like giving the command immediately to the patrolling soldiers. As many times it happens that soldiers are stuck in border areas but they can establish communication with the BSS (base station); sometimes they lose their life and they are unable to inform the BSS (base station) about enemy movement and their visual strategies.

The rest of the paper is organized as follows: Section 3.2 discuss about the literature review; Section 3.3 describes the methodology used in building the soldier strap using IoT; Section 3.4 illustrates the result and discussion followed by Conclusion in Section 3.5.

3.2 LITERATURE REVIEW

Security is one of the main concerns for telecommunication. Numerous encryption methods are used to protect the data. Safe hashing techniques with encryption methods are used for making the data transmission more complete.[4] This algorithm involves asymmetric key generation. There is a patient monitoring system (PMS) which is activated by a wire less personal area network, Chowdhury et al.[5] uses encryption to secure these data. The idea is to encrypt the data with the help of lightweight algorithm of the AES. This kind of the encrypted data are transferred to the MQTT types of the protocol for the hospitals. Savita Sindhu used another version of CSMA/CA and the TDMA protocol for the routing.[6] To record the physiological parameters, such as heart rate for the observation, respiration rate, ECG of the human body, body temperature, and blood pressure, wearable sensors are the kind of portable sensors with nodes are installed inside a wireless body area network (WBAN). Wu et al.[7] suggested hybrid wearable sensors network system that helps workers in the construction industry. To make this hybrid, LPWAN and WBAN, IoT network is required, which uses MQTT protocol for the transmission of the following dataset. IoT brings several challenges like apt infrastructure for control and communications among devices. Bashir et al.[8] used three phases of for security and safety concern schemes, which include key generation, and some kind of the encryption, and also the publication phase through the MQTT and then necessary decryption. Boumaiz et al.[9] investigated using simulation, the impact of some of the particular and adjacent BAN's existence, When the transmitting power differs, on a reference on the study of the BAN's output which is concluded after the simulation. Raza et al.[10] proposed another way of securing data. WBAN uses AES-128 which takes much time to encrypt.

This approach uses scrambling in emergencies based on chaos, taking less time in emergencies. Agha et al.[11] proposed two security suites. The first one focuses on KBS which means the key management and also used with some techniques like Hashing. the second method called KAISC would enhance the communication among the sensors which are used. Kompara et al.[12] used mutual authentication and key agreement which were very efficient for the security concerns in their research which they did after the study of the WBAN experiment methods. Two more new protocols were introduced by Dodangeh et al.[13] to invent the concept of exchange keys between all the sensor nodes with which we would have to watch another topology and also mobile node which is used in a two-tier type of the WBAN topology. A quantized stream encryption scheme has been proposed. PSD and the PAPR analyses, which discovered the quantized form of the chaotic sequences, were performed to check the details.[14] Lightweight encryption algorithm (LEA) seems to be the best for WBAN settings if these kinds of the devices have low memory space and low processing power.[15] It uses the concept of new form which is known as the hybrid encryption to reduce the computational over head. In a new modulation technique, the information which has been applied by the thermal noise which was produced in the connection between its two of the terminals and a kind of loop that recovers the shared information. Rapid changes in technology leads WBAN to a new and advanced version of the smart Healthcare system. Another method to secure the physical layer transmission of data is used by a multihop topology algorithm which is used without any help of the key. WBAN also requires some of the lightweight and more effective way of transmission of the data through a network. Hybrid encryption algorithm (HEA) is also used to secure data over the WBAN and Diffie-Hellman algorithm has been used for key exchange.[16] Another method, which is known as the homomorphic encryption method, has been used for securing the personal dataset of their own patients in the 18 bytes method of the text blocks. By this method, plaintext and ciphertext are obtained.[17] The security as well as safety and also the privacy issues endanger an IoT which seems to be an advancement-based system, which is due to its heterogeneous nature of the part and deprived resources of various IoT devices. Despite LoRaWAN using the AES to protect the data, there are weaknesses in the LoRaWAN system, and many studies have pointed out that.[18] Butun et al.[19] described the weaknesses of LoRaWAN by using the ETSI guidelines and prepared a layout for the concerned system. Miller et al.[20] laid forward some possible attacks on LoRaWAN and suggested some improvements. Yang et al.[21] formulated five main potential attacks on LoRaWAN and these include: a replay of the attack, the plain text recovery, malicious message and the various modification, falsification of the delivery reports, and also the battery exhaustion attack. These researchers also gave some other ways and concept as to how these can be mitigated. To improve thesecurity of the LoRaWAN, some concepts and many new researches required in this field were also done and some methods have been proposed. Tsai et al.[22] developed a security improvement option to prevent the breaking of end-to-end ways of the security, and they also simplified the AES to minimize the encryption and decryption difficulty for LoRaWAN. D box update and dynamic encryption key were also proposed to increase the level of end-to-end security. Sanchez-Iborra et al.[23] explained the short-comings of LoRaWAN regarding the concept and the process of key management done and also proposed a method on the detailed studies of the Ephemeral Diffie-Hellman over the idea of the COSE transactions. Kim et al.[24] discusses some loopholes in the security system of LoRaWAN and updated them for improvement. Oniga et al.[25] implemented a public key feature model after analyzing the security threats of LoRaWAN; this protected the data as each component held a certificate for communicating. The network server is not at all a well-defined in the LoRaWAN, but is used in checking the integrity of the messages and forwarding these messages. Lin et al.[26] introduced blockchain technology and proposed a decentralized LoRaWAN server structure. In addition to this, when the studies had raised a higher concern of security and safety level for the LoRa-AN environment, and it is seen that most of them considered an end-to-end method of the security. Much importance was not given to server-to-server methods of the security. Therefore, we have proposed the S2KG to give a better-secured communication and transmission method for the LoRaWAN servers. In recent studies and research of ML methods, the focus has been about the use of limited computational resources. The Auto-ML idea and the concept of the Neural Architecture Search methods are used to make out resource-efficient edge AI.[27] Various necessary advances in the research have also been made by the surrounding to design of the features of the DL architectures, these are the resource-limited devices. Nikouei et al.[28] put forward a lightweight version of the CNN that can run on edge devices. Several proposals have been considered in the Industrial IoT. For example, in DIoT, the concept of recurrent neural network (RNN) is to be utilized. Wang et al.[29] suggested a control algorithm. Ferdowsi et al.[30] introduced a way of the distributed privacy concept of the network that would preserve IoT invasion based on various types of federated generative adversarial networks concept. In the part where we are performing the decentralized architecture, and the IoT devices monitor their self-data; and at the same time, the neighboring IoT devices to recognize the attacks—both the internal and the external attacks. Meidan et al.[31] suggested N-BaIoT, which is the procedure to detect IoT botnet attacks. In this system, for every device in the network, there is a method known as autoencoder which is trained on the normal concept and considerable area of the traffic data.[29] Bezerra et al.[32] proposed IoTDS as a method to detect IoT botnet attacks in IoT devices which is based lightweight one-class classifiers. Rathore et al.[33] formed a framework for attack detection. Their process was based on the one of the methods known as the semi-supervised learning that is used in the extreme way of learning machines and the fuzzy C-means algorithms. Doshi et al.[34] used many ML techniques to detect DDoS attacks. Nguyen et al.[35] proposed a method for malware detection, which was based on deep RNNs. Kamaraj et al.[36] explained the concept of the Random Forest, or the part of Multilayer Perceptron, and the Discriminant Analysis models can reserve time and also the energy resources on the various edge device during the transmission of data. In addition, it has been shown that K-nearest neighbors are resource-inefficient.

3.3 METHODOLOGY

In this paper, the application of IoT and Advance Encryption Standard-256 (AES-256) have been discussed. Both the technology is the modern way to implement and design our project. The methodology of the project includes the hardware work on the combination of other peripheral devices which are necessary to operate and do the work. In this, we are working on the Arduino as a base microcontroller and using the ESP32 module. Here the ESP32 module is used as an IoT device. This ESP32 will send the data or set up communication through the satellite. In this, we are also using the GPS sensor, heart rate sensor, body temperature sensor, and sound module. The GPS sensor will transmit the coordinates of their latitude and longitude when they are at the patrolling areas and then the location will directly send the data to the base station. After that base will have the exact location of the soldier. The ESP32 module will send all the data to the server. The ESP32 is the Wi-Fi type module and also has a Bluetooth module that will communicate with the server and the other devices that are connected to it. The heart rate sensor will be measuring the heartbeat of the soldier in a fraction of time and send the data to the base. The body temperature will be very effective because at the higher altitude, the soldier's body temperature sometimes goes down or up so they may have medical support and the problems may be solved quickly. There is also a sound sensor module, with that they may communicate with the other soldier in the patrol party and also may send the voice note to the base for urgent help or message. The data are encrypted in the AES encryption because this is the modern encryption technique that is very difficult to decode. Due to this if the enemy gets this band and if they try to open the band forcefully, the band may go to self-destruction mode and the enemy may not get any information from that band. AES is a very safe mode of data transmission that cannot be decoded very easily, and the data will be safe and maybe transmitted safely. The soldier strap is made by me is a prototype that I have implemented through the Internet of Things (IoT) and AES-256. The soldier strap is a band used for safety and security of our soldiers when they are on patrolling or any other operations. This device will ensure their safety and security and also help them in communicating with them from the base station. In this project, the soldier strap is transmitting the data through the Google server and sending the information to the base station. The base station receives the data in a certain time gap set in the particular time frame. The data include the exact latitude and longitude location, heart rate, the body temperature, and voice transmission of the soldier. The data also include the date and time of the data sent and recorded time. The ESP32 module is a basic IoT device used in the project and has a Wi-Fi module as well as bluetooth module. This will help in sending the data to the base station via the Google server. It also has battery support so that the soldier may use it for some days using it continuously and when they return to the base may again plug into the charge. As in this prototype, we have made our own AES encoding and decoding Libraries because the module we have used may not support such complex algorithms, so we have used our own. This device is safe enough because the hardware has been added and then there is a code generated by the server and the code will only be generated once and it is known as the passkey. This passkey once generated for the hardware will be for that particular hardware only. If someone tries to add this hardware to their device, it will not work. So, this AES data encryption is one of the best ways to transmit the data and will also provide safety and security. Table 3.1 presents the purpose and function of sensors.

TABLE 3.1 Shows the Purpose and Function of Sensors.

S. No	Purpose	Function
1.	Tracking	The strap enables the tracking of the location using the method of defense study of mapping so that they may track the soldier with their band or with the help of the soldier strap in which the GPS device is fixed when they are on patrolling. The GPS will send the exact latitude and longitude of the particular soldier. Hence it is convenient for the BSS (base station) to send help to that soldier
2.	ESP32	They may transfer the information in the Advance Encryption Standard (AES) data encryption so that the data will be secure and can be only decoded by the BSS (base station) or the headquarters. The data will be communicated through the satellite medium; it is an easy way to contact the soldier to BSS (base station) and vice versa. The data in AES is so advanced version in that it is very difficult to decode the information for our enemy
3.	Blood pressure	This device has blood pressure equipment that will check the BP and then send the data to each and base too
4.	Camera	The strap also has a camera that may click pictures that the soldier will send to the base. This will help the base to plan a strategy and send backup for their soldiers. With the help of images, the base may know the exact locations of the enemy and with the help of the GPS coordination, they may also help their soldier through artillery and through their air force immediately. The image will be sent in encrypted form; and the enemy cannot interpret the data immediately
5.	Temperature	The strap also has a device that may find the body temperature and keep the data within the device. This body temperature is necessary because in the J and K areas, the temperature is very low (in minus), so when the soldier moves in the forward location, we may track the temperature of the soldier so that if any kind of unusual things are noticed, then we can contact to their associates for help and support
6.	Locking system	The strap will have a special locking part which will be done by the base before they start their patrolling or any kind of operation. In case if the strap is been broken forcefully, then all the last data which have been there in the device will be sent to the base and the device will self-destroy automatically. This is majorly done for the safety purpose because in case if any of the enemies trap our soldier and try to torture them and want to extract the data from the device, and if they forcefully try to open, then the soldier strap automatically goes on the self-destroy mode and the last data will be sent to the base and after that the base may take certain action to save their soldier

TABLE 3.1 *(Continued)*

S. No	Purpose	Function
7.	SOS1	The SOS (Save Our Soul) is the button for the emergency purpose of the soldier. The strap may have a button in case if he presses the button, then a signal will be generated and will transmit to all his colleagues and the base signaling that the soldier is in danger, and by the GPS location, they may track the device and the soldier too. This SOS1 is basically useful if there is an enemy attack or any unusual situation happens at the Line of Control, then the soldier may ask for the help from the base and from the colleagues
8.	SOS2	Another SOS button will be there for any kind of medical help so that they may provide the medical help to that soldier
9.	Bluetooth module	This Bluetooth module is fixed in the soldier strap, and it is used for the communication of collogues who are present at that time of patrolling. This will also be useful in case if any soldiers are having communication with the satellite then they may transfer to each other and if any one of them gets the signal then they may directly upload or transmit to the base. By this, they can also communicate with each other for any kind of help and actions
10.	Sound module	This sound module is fixed in the soldier strap to contact each other and can talk and share their information's among them. The sound module is also used for sending the voice note to the base
11.	Heart rate sensor	This heart rate sensor has also been fixed in the soldier strap which will measure the reading of the heartbeat of the soldier and provide information about the medical fitness and the detailed report of the soldier conditions
12.	Danger alarm switch	This is the alarm switch which will be used by the soldiers when they are facing any kind of sudden or unpredictable situation so that by pressing the alarm button they can inform the base about the warning with the help of this base deference voltage system applied in this soldier strap.
13.	AES data encryption	AES is one of the finest methods used for transmitting the data through the network or satellite. The concept has consisted of basically various way and there are basically three major block ciphers methods: AES-128, AES-192, and AES-256. AES-128 uses a 128-bit key length to encrypt and also to de-crypt a block of messages; the method has also another version which is AES-192. It is used for 192-bit key length methods, and after this the third concept of the AES is AES-256. This is a method in which the data set is transmitted using the 256-bit key length for the dataset which may encrypt and de-crypt the following messages. In every cipher encryption and decryption method, the data are sent in blocks of 128 bits which is the basic step and also by using this crypto-graphic concept of the keys of 128, 192, and 256 bits, respectively. This is the safest and can't be cracked by the enemies so easily when the data is transmitted. And can be used as the best way to be used by the soldiers for transferring the data from the border or from the patrolling areas

3.4 RESULT AND DISCUSSION

Figure 3.1 depicted the designed hardware of the soldier strap on PCB board. The ESP32 module and GPS module with antenna are attached, and in the center, it is a body temperature sensor; there is a green color light device which is a heart rate sensor and there is a sound module. The data are received from the hardware and then sent through the Google server. This prototype is through the AC supply but for the use of the soldier, it will be the connected battery.

FIGURE 3.1 Hardware design of the soldier strap.

Figure 3.2 shows the output received at the base station in the form of a table. This is the proper format of the result received and it is sent through the Google server in the form of a table. As in my prototype of the project, I have received the data in this particular format where the GPS location is in the exact format of latitude and longitude like the same way as defense personal use to locate any of the geographical areas in the map, and then the body temperature of the soldiers is measured in Fahrenheit and the heart rate of the soldiers is displayed along with the date and time when the data has been transmitted to the base and of which soldier along with their device no. Therefore, in the database, we may locate the information of which more armed personnel have done the patrolling up to which parts of the border, and which part of the border are left unchecked so that others might visit those areas to check them out. Also, the search troops with get help for the communication and help them. And the base station will have the complete records and also medical records of all the soldiers present at the border at that time. If any problems occur, then the soldier will get the treatments to make his body fit from those issues. Here all the data received will be safe and secure because by using the Advance Encryption Standard 256, the data privacy will not be breached and the data transmitted by the soldiers will be safe. The data can be processed or read only by the base stations. Therefore, the Advance Encryption Standard 256 is the best way for an algorithm to transmit the data. Because it is the concern with the safety and security of the nation and our soldier's life too.

	A	B	C	D	E
1	Time	Latitude	Longitude	Body Temperatu	Heart Beat rate
2	00:00:07-14/12/	28.58	77.08	73.89	0
3	00:00:26-14/12/	28.58	77.08	73.81	0
4	00:00:44-14/12/	28.58	77.08	73.6	0
5	00:00:59-14/12/	28.58	77.08	73.45	0
6	00:01:14-14/12/	28.58	77.08	73.31	0
7	00:01:29-14/12/	28.58	77.08	73.27	0
8	00:01:44-14/12/	28.58	77.08	73.13	0
9	00:02:03-14/12/	28.58	77.08	72.77	0
10	00:02:18-14/12/	28.58	77.08	72.41	0
11	00:02:33-14/12/	28.58	77.08	72.23	0
12	00:02:49-14/12/	28.58	77.08	72.16	0
13	00:03:04-14/12/	28.58	77.08	72.3	0
14	00:03:19-14/12/	28.58	77.08	72.34	0
15	00:03:35-14/12/	28.58	77.08	72.41	0
16	00:03:50-14/12/	28.58	77.08	72.41	0
17	00:04:04-14/12/	28.58	77.08	72.52	0
18	00:04:19-14/12/	28.58	77.08	72.63	0
19	00:04:33-14/12/	28.58	77.08	72.59	0
20	00:04:48-14/12/	28.58	77.08	72.77	0
21	00:05:03-14/12/	28.58	77.08	72.84	0
22	00:05:17-14/12/	28.58	77.08	72.91	0
23	00:05:33-14/12/	28.58	77.08	72.99	0
24	00:05:48-14/12/	28.58	77.08	72.99	0
25	00:06:04-14/12/	28.58	77.08	73.2	0
26	00:06:08-14/12/	28.58	77.08	73.17	0
27	00:06:23-14/12/	28.58	77.08	73.2	0
28	00:06:41-14/12/	28.58	77.08	73.2	0
29	00:06:56-14/12/	28.58	77.08	73.31	0
30	00:07:13-14/12/	28.58	77.08	73.31	0
31	00:07:31-14/12/	28.58	77.08	73.42	0

FIGURE 3.2 Outputs received at the base station.

Figure 3.3 shows a GPS module in which the GPS sensor is embedded along with its antenna. This antenna helps us to give the accurate direction of the location. Hence, we have attached this GPS module along with its antenna to the soldier's strap.

FIGURE 3.3 GPS sensor is embedded along with its antenna.

Figure 3.4 shows the sound module which is embedded with mic and speaker. As in this picture, the Red color button is used for recording the voice of the soldier, and after that, there are two buttons. The first button is SOS1 which is used to send the data to his associates who are there on the patrolling time and the second button is the SOS2 button which will directly send the data to the base station. The base station will listen to the message passed by the soldier and will take further action. The speaker will be helpful to listen to the audio sent by any of them means either from the base station or from the patrolling parties.

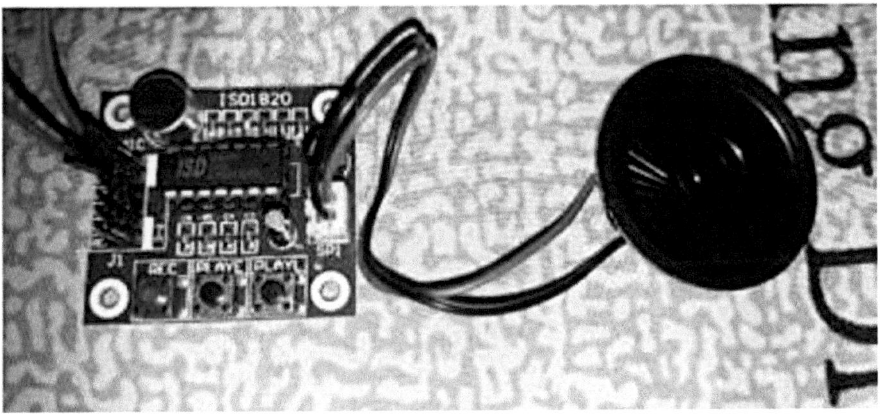

FIGURE 3.4 Sound Module in which is embedded with mic and speaker.

Figure 3.5 depicts the heart rate sensor. This sensor is used to measure or get the exact reading of the heart rate of a particular soldier on the patrolling. This module consists of three pins in which the first one is for the ground, the second one is for the +5 volt, and the third pin is for the pulse counting. And these signals are sent to the microcontrollers and the data are then further processed. This sensor blinks by green light and after that, if we put a finger on it, then the color changes to red.

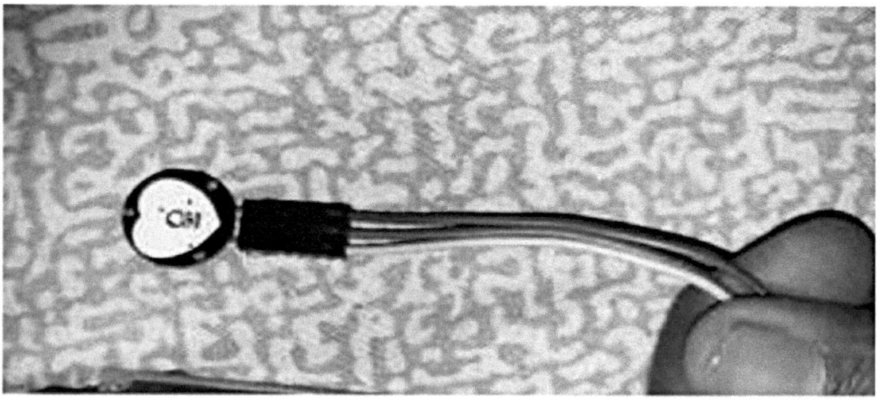

FIGURE 3.5 Heart rate sensor.

Figure 3.6 depicted ESP32 module. This is used as a microcontroller that is connected to all the peripheral modules or hardware used in this soldier strap. The ESP32 module is also used for transmitting the data or uploading the data over the server as it works as an IoT module. It also contains a Bluetooth module. It contains the analog pins as well as digital

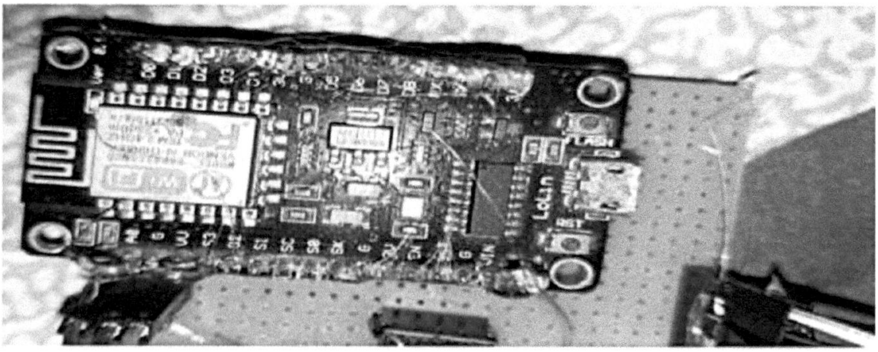

FIGURE 3.6 ESP32 module.

pins. We have connected the hardware according to the needs and requirements. All the peripheral devices or hardware modules are embedded over the PCB designing board and the connection is done to connect all the pins by using a female jack of pin holders. This is the main device used for communication and transferring the data.

3.5 CONCLUSION

Based on the expected result that has been tested using designated tools and platforms. In addition, this project has fulfilled its objective. The device aids in tracking the exact position of our Indian soldier at any point or any interval of time by using the GPS system. Since the main focus of this project is the soldier's health parameters, such as heart rate of the soldier, which are particularly important at the higher hilly areas, and the body temperature, which is continuously measured; if it fluctuates in any case, then we can inform to the other colleagues of the patrolling parties and transmitted all the related data through wireless systems to the base station or the control room where it can be used to provide the exact coordinates using GSM system to track and provide or take the necessary steps. In case of demise of the Indian soldier, the processor will automatically detect and fetch all the data from the device and base station can check and then it will be helpful to track the pulse rate and also to track their exact location of the dead Indian soldier and further can send any backup to retrieve the bodies using the GPS system module, and then they can also communicate to the nearest Indian military base stations and ask for further permission or orders to do the next process by utilizing the GSM. This concept of the soldier strap is also very useful, and the information it contains can be used to strategically plan how many more soldiers should be deployed to replace the soldiers who have died in battle or during any of the defense exercise to train our soldiers, As in any of the emergency situations, the Indian soldiers can also communicate directly for the further necessary order by raveling their exact situations to the base station using a danger alarm switch. The device is used to monitor and provide the health status of the soldier continuously and also the exact updates of the border and the patrolling areas.

The future scope is to implement in the Indian defense, therefore, this can be used for the safety and security of the soldier so that the base may find their location in any of the adverse cases. This will also empower the potential of soldiers and provide a morale boost so that they may have complete trust in their safety and security. This will also bolster the strength of the Indian soldier when they are at the international border or any of the missions. This will be helpful to track them at any point in time and may also pass any of the important messages to them. As this device is at an affordable cost for the Indian Government. Hence, they may provide to the soldiers for their betterment and safety and security of the country. As many times it has been noticed that the enemy infiltration are seen at the border areas and the soldiers cannot get help from the base because they cannot communicate and if they try to contact them through the radio frequency, the enemy will try to listen to their message or just will try to hack the message. By this device, it will be helpful for our soldiers to pass the message over the base station in any adverse conditions. in the future, the soldier strap will be more helpful for our Indian defense soldiers to operate any of the operations in any of the areas. This soldier strap, the soldier and the base station will also check all the forward post and the areas, and can also have a survey of the areas where the chances of the disturbances are maximum and may keep an eye on that particular area very keenly. In this, we may also add other essential equipment's in the strap which will provide more variety of the data from the soldiers, such as cameras, step-counting devices, and night vision types of equipment. In the future, if the base will get the information on accurate time then will help our soldiers by Air Rescue medium or through ground attack vehicles like Bofors gun or artillery gun and many more other help to the soldiers so that they can perform their operation properly.

KEYWORDS

- **Internet of things**
- **security**
- **SOS signals**
- **AES data encryption**
- **GPS**

REFERENCES

1. Sammoud, A.; Chalouf, M. A.; Hamdi, O.; Bouallegue, A.; Montavont, N. A New Biometrics-Based Key Establishment Protocol in WBAN: Energy Efficiency and Security Robustness Analysis. *Comput. Secur.* **2020,** *96,* Art. no. 101838.
2. Wang, J.; Han, K.; Fan, S.; Zhang, Y.; Tan, H.; Jeon, G.; Pang, Y.; Lin, J. A Logistic Mapping-Based Encryption Scheme for Wireless Body Area Networks. *Future Gener. Comput. Syst.* **2020,** *110,* 57–67.
3. You, I.; Kwon, S.; Choudhary, G.; Sharma, V.; Seo, J. An Enhanced LoRaWAN Security Protocol for Privacy Preservation in IoT with a Case Study on a Smart Factory-Enabled Parking System. *Sensors* **2018,** *18* (6), 1888.
4. Anwar, M.; Abdullah, A. H.; Butt, R. A.; Ashraf, M. W.; Qureshi, K. N.; Ullah, F. Securing Data Communication in Wireless Body Area Networks Using Digital Signatures. *Tech. J.* **2018,** *23* (2), 50–55.
5. Chowdhury, F. S.; Istiaque, A.; Mahmud, A.; Miskat, M. In *An Implementation of a Lightweight end-to-end Secured Communication System for Patient Monitoring System*, Proc. Emerg. Trends Electron. Devices Comput. Techn. (EDCT); At Guru Nanak Institute Technology: Kolkata, India, 2018; pp 1–5.
6. Sindhu, S.; Vashisth, S.; Chakarvarti, S. K. A Review on Wireless Body Area Network (WBAN) for Health Monitoring System: Implantation Protocol. *Commun. Appl. Electron.* 2016, *4* (7), 1–5.
7. Wu, F.; Wu, T.; Yuce, M. R. An Internet-of-Things (IoT) Network System for Connected Safety and Health Monitoring Applications. *MDPI Sens.* **2018,** *19* (1), 1–21.
8. A Bashir and Mir, A. H. Securing Communication in MQTT Enabled Internet of Things with lightweight Security Protocol. EAI Endorsed Trans. *Internet Things* **2018,** *3* (12), 1–6.
9. Boumaiz, M.; El Ghazi, M.; Bouayad, A.; Fattah, M.; El Bekkali, M.; Mazer, S. In *The Impact of Transmission Power on the Performance of a WBAN Prone to Mutual Interference*, Proc. Int. Conf. Syst. Collaboration Big Data, Internet Things Secur. (SysCoBIoTS); Casablanca, Morocco, 2019; pp 1–4.
10. Raza, S. F.; Naveen, C.; Satpute, V. R.; Keskar, A. G. In *A Proficient Chaos Based Security Algorithm for Emergency Response in WBAN Sys*tem, Proc. IEEE Students' Technol. Symp. (TechSym); Nagpur, India, 2016; pp 18–23.
11. Agha, D. S.; Khan, F. H.; Shams, R. A Secure Crypto Base Authentication and Communication Suite in Wireless Body Area Network (WBAN) for IoT Applications. *Wireless Pers. Commun.* **2018,** *103,* 2877–2890.
12. Kompara, M.; Islam, S. H.; Hölbl, M. A Robust and Efficient Mutual Authentication and key Agreement Scheme with Untraceability for WBANs. *Comput. Netw.* **2019,** *148,* 196–213.
13. Dodangeh, P.; Jahangir, A. H. A Biometric Security Scheme for Wireless Body Area Networks. *J. Inf. Secur. Appl.* **2018,** *41,* 62–74.
14. Alshamsi, A. Z.; Barka, E. S. In *Implementation of Energy Efficient/Lightweight Encryption Algorithm for Wireless Body Area Networks*, Proc. Int. Conf. Informat., Health Technol. (ICIHT); Riyadh, Saudi Arabia, 2017; pp 21–23.

15. Vora, J.; Nayyar, A.; Tanwar, S.; Tyagi, S.; Kumar, N.; Obaidat, M. S.; Rodrigues, J. J. P. C. In *BHEEM: A Blockchain-Based Framework for Securing Electronic Health Records*, Proc. IEEE Globecom Workshops (GC Wkshps); United Arab Emirates: Abu Dhabi, 2018; pp 9–13.
16. Farooq, S.; Prashar, D.; Jyoti, K. Hybrid Encryption Algorithm in Wireless Body Area Networks (WBAN). In *Intelligent Communication, Control and Devices*; Singh, R., Choudhury, S., Gehlot, A., Eds.; Springer Nature: Singapore, 2018; p 10.
17. Hathaliya, J. J.; Tanwar, S. An Exhaustive Survey on Security and Privacy Issues in Healthcare 4.0. *Comput. Commun.* **2020,** *153,* 311–335.
18. Gritti, C.; Önen, M.; Molva, R.; Susilo, W.; Plantard, T. Device Identification and Personal Data Attestation in Networks. *J. Wireless Mobile Netw. Ubiquitous Comput. Dependable Appl.* **2018,** *9* (4), 1–25.
19. Butun, I.; Pereira, N.; Gidlund, M. Security Risk Analysis of LoRaWAN and Future Directions. *Future Internet* **2019,** *11* (1), 3.
20. Miller, R. *LoRa Security–Building a Secure LoRa Solution*; MWR Labs: New York, NY, USA, White Paper, 2016.
21. Yang, X.; Karampatzakis, E.; Doerr, C.; Kuipers, F. In *Security Vulnerabilities in LoRaWAN*, Proc. IEEE/ACM 3rd Int. Conf. Internet-Things Design Implement (IoTDI); Orlando, FL, USA, 2018; pp 129–140.
22. Tsai, K. L.; Leu, F. Y.; Yang, L. C.; Li, C.; Liu, J. Y. In *LoRaWAN Network Server Session Keys Establish Method with the Assistance of Join Server, Proc. MobiSec; Taicung*, Taiwan, 2019; pp 1–6.
23. Sanchez-Iborra, R.; Sánchez-Gómez, J.; Pérez, S.; Fernández, P.; Santa, J.; Hernández-Ramos, J.; Skarmeta, A. Enhancing LoRaWAN Security Through a Lightweight and Authenticated key Management Approach. *Sensors* **2018,** *18* (6), 1833.
24. Kim, J.; Song, J. A Dual Key-Based Activation Scheme for Secure LoRaWAN. *Wireless Commun. Mobile Comput.* **2017,** *2017,* 6590713.
25. Oniga, B.; Dadarlat, V.; De Poorter, E.; Munteanu, A. In *A Secure LoRaWAN Sensor Network Architecture*, Proc. IEEE Sensors; Glasgow, U.K., 2017; pp 1–3.
26. Lin, J.; Shen, Z.; Miao, C.; Liu, S. Using Blockchain to Build Trusted LoRaWAN Sharing Server. *Int. J. Crowd Sci.* **2017,** *1* (3), 270–280.
27. He, Y.; Lin, J.; Liu, Z.; Wang, H.; Li, L.-J.; Han, S. In *AMC: Automl for Model Compression and Acceleration on Mobile Devices*, Proc. Eur. Conf. Comput. Vis., Cham; Switzerland: Springer, 2018; pp 815–832.
28. Nikouei, S. Y.; Chen, Y.; Song, S.; Xu, R.; Choi, B.-Y.; Faughnan, T. In *Smart Surveillance as an Edge Network Service: From Harr-Cascade, SVM to a Lightweight CNN*; Proc. IEEE 4th Int. Conf. Collaboration Internet Comput. (CIC); 2018; pp 256–265.
29. Wang, S.; Tuor, T.; Salonidis, T.; Leung, K. K.; Makaya, C.; He, T.; Chan, K. Adaptive Federated Learning in Resource Constrained Edge Computing Systems. *IEEE J. Sel. Areas Commun.* **2019,** *37* (6), 1205–1221.
30. Ferdowsi, A.; Saad, W. In *Generative Adversarial Networks for Distributed Intrusion Detection in the Internet of Things*, Proc. IEEE Global Commun. Conf. (GLOBECOM); 2019; pp 1–6.

31. Meidan, Y.; Bohadana, M.; Mathov, Y.; Mirsky, Y.; Shabtai, A.; Breitenbacher, D.; Elovici, Y. N-BaIot—Network-Based Detection of IoT Botnet Attacks Using Deep Autoencoders. *IEEE Pervasive Comput.* **2018,** *17* (3), 12–22.
32. Bezerra, V. H.; da Costa, V.; Barbon, Jr., S.; Miani, R. S.; Zarpelão, B. B. IoTDS: A One-Class Classification Approach to Detect Botnets in Internet of Things Devices. *Sensors* **2019,** *19* (14), 3188.
33. Rathore, S.; Park, J. H. Semi-Supervised Learning Based Distributed Attack Detection Framework for IoT. *Appl. Soft Comput.* **2018,** *72,* 79–89.
34. Doshi, R.; Apthorpe, N.; Feamster, N. In *Machine Learning DDoS Detection for Consumer Internet of Things Devices*, Proc. IEEE Secur. Privacy Workshops (SPW); 2018; pp 29–35.
35. Nguyen, T. D.; Marchal, S.; Miettinen, M.; Fereidooni, H.; Asokan, N.; Sadeghi, A.-R. In *DÏoT: A Federated Self-Learning Anomaly Detection System for IoT*, Proc. IEEE 39th Int. Conf. Distrib. Comput. Syst. (ICDCS); 2019; pp 756–767.
36. Kamaraj, K.; Dezfouli, B.; Liu, Y. In *Edge Mining on IoT Devices Using Anomaly Detection*, Proc. Asia–Pacific Signal Inf. Process. Assoc. Annu. Summit Conf.; 2019; pp 33–40.
37. Umar, M.; Sabir, Z.; Raja, M. A. Z.; Gupta, M.; Le, D. N., et al. Computational Intelligent Paradigms to Solve the Nonlinear SIR System for Spreading Infection and Treatment Using Levenberg–Marquardt Backpropagation. *Symmetry* **2021,** *13* (4), 618–636.
38. Gupta, M.; Lechner J.; Agarwal, B. Performance Analysis of Kalman Filter in Computed Tomography Thorax for Image Denoising. *Recent Adv. Comput. Sci. Commun.* **2020,** *13* (6), 1199–1212.
39. Gupta, M.; Taneja, H.; Chand, L.; Goyal, V. Enhancement and Analysis in MRI Image Denoising for Different Filtering Techniques. *J. Stat. Manag. Syst.* **2018,** *21* (4), 561–568.
40. Nisar, K.; Sabir, Z.; Raja, M. A. Z.; Ibrahim, A. A. A.; Rodrigues, J. J., et al. Evolutionary Integrated Heuristic with Gudermannian Neural Networks for Second Kind of Lane–Emden Nonlinear Singular Models. *Appl. Sci.* **2021,** *11* (11), 4725–4740.
41. Gupta, M.; Taneja, H.; Chand, L. Performance Enhancement and Analysis of Filters in Ultrasound Image Denoising. *Procedia Comput. Sci.* **2018,** *132* (1), 643–652.
42. Kumar, A.; Chakravarty, S.; Gupta, M.; Baig, I.; Albreem, M. A. Implementation of Mathematical Morphology Technique in Binary and Grayscale Image. *Adv. Concepts Image Proc. Pattern Recognit.* Ed. **2022,** *1,* 203–212.
43. Kumar, S.; Kumar, P.; Gupta, M.; Nagawat, A. K. Performance Comparison of Median and Wiener Filter in Image De-Noising. *Int. J. Comput. Appl.* **2010,** *12* (4), 27–31.
44. Umar, M.; Sabir, Z.; Raja, M. A. Z.; Shoaib, M.; Gupta, M., et al. A Stochastic Intelligent Computing with Neuro-Evolution Heuristics for Nonlinear SITR System of Novel COVID-19 Dynamics. *Symmetry* **2020,** *12* (10), 1628–1646
45. Gupta, M.; Upadhyay, S.; Nagawat, A. K. Camera Calibration Technique Using Tsai's Algorithm. *Int. J. Enterprise Comput. Bus. Syst.* **2011,** *34* (2), 190–213.
46. Umar, M.; Sabir, Z.; Raja, M. A. Z.; Gupta, M.; Le, D. N., et al. Computational Intelligent Paradigms to Solve the Nonlinear SIR System for Spreading Infection

and Treatment Using Levenberg–Marquardt backpropagation. *Symmetry* **2021,** *13* (4), 618–635.

47. Nisar, K.; Sabir, Z.; Raja, M. A. Z.; Ibrahim, A. A. A.; Rodrigues, J. J. P. C., et al. Evolutionary Integrated Heuristic with Gudermannian Neural Networks for Second kind of Lane–Emden Nonlinear Singular Model. *Appl. Sci.* **2021,** *11* (11), 4725–4747.
48. Gupta, M.; Lechner, J.; Agarwal, B. Performance analysis of Kalman Filter in Computed Tomography Thorax for Image Denoising. *Recent Adv. Comput. Sci. Commun.* **2020,** *13* (6), 1199–1212.
49. Gupta, M.; Taneja, H.; Chand, L.; Goyal, V. Enhancement and Analysis in MRI Image Denoising for Different Filtering Techniques. *J. Stat. Manag. Syst.* **2018,** *21* (4), 561–568.
50. Haque, M. R.; Tan, S. C.; Yusoff, Z.; Nisar, K.; Lee, C. K., et al. Automated Controller Placement for Software-Defined Networks to Resist DDoS Attacks. *Comput. Mater. Continua* **2021,** *68* (3), 3147–3165.
51. Gupta, M.; Chand, L.; Pareek, M. Power Preservation in OFDM Using Selected Mapping (SLM). *J. Stat. Manag. Syst.* **2019,** *22* (4), 763–771.
52. Nisar, K.; Sabir, Z.; Raja, M. A. Z.; Ibrahim, A. A. A.; Rodrigues, J. J. P. C., et al. Artificial Neural Networks to Solve the Singular Model with Neumann–Robin, Dirichlet and Neumann Boundary Conditions. *Sensors* **2021,** *21* (19), 6498–6521.

CHAPTER 4

BIOSENSING: A PROMISING APPROACH FOR BIOANALYSIS USING IoT

MAYA DATT JOSHI[1] and AMIT KUMAR SINGH[2]

[1]*Department of Biotechnology, GLA University, Mathura, India*

[2]*Biomedical Engineering Department, VSB Engineering College, Karur, India*

ABSTRACT

Biosensing is a technology-based approach to the analysis of biomolecules. This approach is tremendously used in every area of biological research, including the analysis of pollutants or toxic components in environmental samples like food, soil, or water. Biosensing is used in drug delivery systems through the use of biomedical devices, including biomedical microelectromechanical systems. Enzyme-based electrochemical biosensors are frequently used in medical sciences to analyze the health parameters of the human body. Various biosensor-based nanodevices, including carbon nanotubes (CNTs), are used in drug delivery to treat diseases like cancer, cardiovascular disorders, diabetes mellitus, and others. In addition, biosensing plays an essential role in designing detection kits based on immunological assays for several pathogenic and infectious diseases, including the recent coronavirus disease (COVID-19) infection. Various biosensing systems have combinations of biological and microelectronic devices using IoT technology for real-time detection of target analytes and are presently being used for various applications. Nowadays, there is more

IoT-Enabled Healthcare Systems: Applications, Benefits, and Challenges
Yatindra Kumar, Shashi Kant Verma, Archana Singh, Krishna Kumar, & Manoj Gupta (Eds.)

focus on developing cost-effective IoT-based biosensing devices that can be used effectively with the highest sensitivity and provide more accurate output within less time, which is the prerequisite of the present world for commercialization. This chapter has discussed the different applications, pros, and challenges in the IoT biosensor (IoTB) field.

4.1 INTRODUCTION

Primarily, it is essential to understand the concepts of sensors before going into the details of biosensors. Sensors detect changes in physical quantities such as temperature in a region, surrounding humidity, water flow rate, light intensity, and others and convert them into quantities that can be measured and studied. At the same time, a biosensor transforms a biological signal into an electrical signal, which is a more usable form of energy.[16] Natural materials or elements such as bacteria, cells, tissues, acids, enzymes, and others can be biological processes. The biosensor consists of a detecting element of biological activity and a transducer that translates the data into the form of an electrical signal. It also consists of a signal conditioning unit, a processing unit, and a display unit on an electronic circuit.[49]

Biosensors are widely employed in today's technological era and have several applications in various research fields. Disease detection, environmental monitoring, drug discovery, agriculture, pharmaceutical development, and many more fields of research use biosensors. The current chapter concentrates on biosensors' essential components, concepts, operations, and applications in different parts of human life. Biosensing technology is widely employed, and biosensing devices are desperately needed to solve various societal development and health problems. Biosensors are widely used for biomedical diagnostics and applications based on illness monitoring, point-of-care treatment, food safety, drug discovery, environment monitoring, and biomedical research. A comprehensive overview of biosensors and biosensing technologies, including a brief historical review, essential advancements in the field, and examples of the wide range of biomolecular sensing techniques and nanotechnological approaches now accessible, is shown in this chapter.

The use of fitness and well-being monitoring technology has been regularly growing, and such systems can now-a-days be observed in smart homes, age-friendly offices, public spaces, and other places. They install a considerable variety of off-the-shelf smart sensors and clinical devices to assist purposeful, physiological, and behavioral monitoring, address social interplay aspects of everyday existence, and raise awareness both in specific fitness-related situations and in supporting the greater general intention of consolation, well-being, and greatness of existence. However, there are nonetheless numerous technological (interoperability, expandability, and others) and societal (cost privacy and others) demanding situations to be addressed earlier than smart biosensor structures are extensively followed.[42] Smart biosensors can help those who require medical assistance or care and older people experiencing a gradual loss of physical, cognitive, and other skills as they age. It can also benefit those who operate in dangerous locations, undertake stressful duties, or conduct tasks that put their health at risk. Smart systems can create a reactive living and working environment by continuously monitoring biosignals and combining them with environmental sensing to deliver suitable and timely recommendations, act preventively, and decrease health concerns. The chapter discusses the use of IoT technology in biosensor applications and how the same biosensor can benefit human society more in terms of healthcare and other monitoring requirements.

4.2 PRINCIPLE OF AN IoTB

The different principles used in biosensor development and their classification are shown in Figure 4.1. Among many bioreceptor enzymes, the essential biological material is determined by using the electroenzymatic method, a chemical process that uses a transducer to transform enzymes into matching electrical signals (typically electricity) to detect the analyte concentration. One of the most frequently used oxidations of the enzyme is the biological response. The biological material's pH is altered by oxidation, which functions as a catalyst. The catalyst capacity of the enzyme is directly affected by changes in pH, which is, once again, related to the enzyme being examined.[12] The transducer's current output directly measures the value of the analyte concentration. Mostly, current output is converted to voltage form and then displayed. After that, the microcontroller sends that data to the IoT device for internet-based communication.

4.3 COMPONENTS OF IoTB

The components of the IoT-based biosensor are shown in Figure 4.1. The six main parts that make up the IoT are samples, bioreceptors, transducers, signal conditioning circuits, signal processing circuits,[32] and the IoT itself. The different samples that a biosensor can detect for the analyte are also shown in the figure. These contain human biological samples, food samples, environmental samples, drugs, diseases, and other biological samples.[32] The two parts of a biosensor consist of a biological component to detect the presence and concentration values of the required parameters and a transducer device, respectively. First, the sample can pass through a membrane, allowing selection while interfering molecules are kept outside the barrier. After that, the sample interacts with the biological sensor to produce a product, which could be an electric current/charge, heat, gas, or an appropriate chemical. Finally, after passing through another membrane, the product reaches the transducer. The bioreceptors used to convert the analyte concentration to a signal understandable by the transducer are enzymes, nucleic acids, bacteria, antibodies, cells, and tissues.[32] The transducer selection is based on the output of the bioreceptors, which are therefore classified as electrochemical, optical, calorimetric, mass, magnetic, and others.[32] The signal conditioning circuit must filter and increase the signal's amplitude up to the range acceptable by the analog-to-digital converter (ADC). For this purpose, amplifiers, filters, multiplexers, ADCs, and linearizers are used. After that, the signal reaches the microcontroller, where the data processing is done to convert the voltage into the analyte concentration

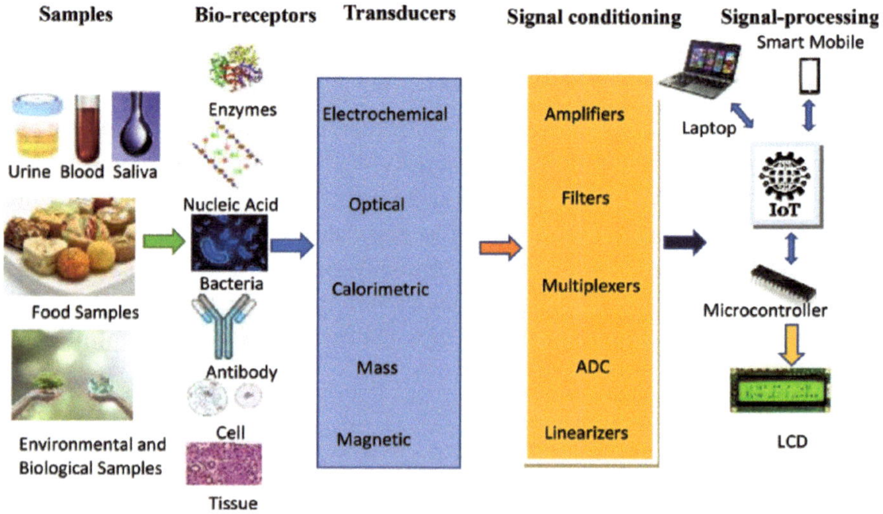

FIGURE 4.1 Schematic diagram of IoT-based biosensor.

based on the calibration equation found during the development process of the biosensor, and then, it displays the data onto the liquid crystal display (LCD). The data can be sent remotely for the requirements of another user/entity with IoT's help in real-time.

There are a variety of commercial biosensors that are being utilized for numerous applications in society. A list of the different types of commercial biosensors is provided in Table 4.1. The table is created based on the transducer used in the biosensor development. These are amperometric immunosensors, potentiometric immunosensors, impedimetric immunosensors, conductometric immunosensors, optical biosensors, piezoelectric, and magnetoelectric biosensors. Table 4.1 also shows the applications of commercial biosensors and the examples that were also discussed in Table 4.1. More than 500 businesses are now developing biosensors and bioelectronics throughout the world. The focus of these businesses is on the creation and marketing of these biosensors. In contrast, others play a vital role in supplying biosensor raw materials, reagents, and equipment. These firms' focus is on biosensor technologies invented more than a decade ago. Some are developing newer technologies, but they are upgrading current ones to bring them to market and interfacing them with IoT. Some firms manufacture biosensor parts, while others assemble the components and market them; still, others manufacture biosensor parts and entire sets. As a result, following the criteria to award credit for the invention of commercial biosensors for all of them is challenging.

Commercial biosensors are split into two types, and these are based on laboratory or portable/field devices. The blood glucose monitor for diabetics, which uses electrochemical transduction technology, is the most successful handheld biosensor. Many firms manufacture commercial blood glucose meters. However, an optical detection method looks to be more financially viable in terms of laboratory-based gear. Companies like Affymetrix and Agilent have created commercial microarray optical detectors and scanners for proteomic and genomic investigation. Many laboratories and universities have successfully used optical sensors based on Surface Plasmon Resonance (SPR) detection.

Several businesses are exploring implanted biosensors, although they mainly focus on monitoring blood glucose levels. Medtronic Minimed, the company's diabetes treatment division, created a tiny sensor based on an enzyme that may be embedded below the skin for up to three days.

TABLE 4.1 A Few Examples of Commercial Biosensors and Their Applications.

S. No.	Types of commercial biosensors	Application of commercial biosensors	Examples of biosensors
1	Amperometric Immunosensor	Detection of cancer biomarkers and illness	Nanogold/TH-fGRNS-AuE
2	Potentiometric Immunosensor	Nanoparticle labeling and ion-selective electrodes	CdSe nanoparticles were used on secondary antibodies in a sandwich immunoassay to obtain a protein biosensor
3	Impedimetric Immunosensor	Detection of microorganisms in the environment or food materials.	Antibody-based sensors, nucleic acid-based sensors, bacteriophage-based sensors, and lectin-based sensors
4	Conductometric Immunosensor	Detection of different pesticides, herbicides, and heavy metal ions	To detect toxic compounds such as copper ions, atrazine, and diuron using a tyrosinase biosensor
5	Optical Biosensors	Detection of biomolecular interactions	BIAcore™ and 3D-printed microfluidic chips coupled with silicon photomultipliers
6	Piezoelectric Biosensor	Detection of oscillation exchange due to mass change in the piezoelectric crystal	Quartz crystal microbalance (QCM) platform
7	Magnetoelectric Biosensor	The magnetic particles bind with the target molecule in the biological assay, creating the magnetic field, which is utilized further	Giant magnetoresistive sensors and superconducting quantum interference devices

VeriChip, on the other hand, is working on an implanted microprocessor. While this is a step in the right direction, these technologies nonetheless have limits. It is worth noting that many multinational businesses are pioneering the biosensor industry sector. According to studies, Roche Diagnostic, MediSense, and Bayer are the industry leaders in handheld "meter" style devices with disposable one-shot electrodes (Newmann et al., 2004).[51] Abbott's recent merger of Therasense and i-STAT, on the other hand, has significantly enhanced its position among the world's leading biosensor companies. For the detection and evaluation of biomolecular interactions, commercially accessible immune-sensing systems by IAsys and BIAcoreTM have research laboratory-based research to market.[39]

4.4 DIFFERENT TYPES OF BIOSENSORS

The two categories of biosensors that employ a biological element in the analysis are those that use a transduction mechanism. Enzymes, DNA, antibodies, bacteria, cell receptors, tissues, and biorecognition or biological elements are the most often used biorecognition elements.[48] In addition, biosensors' most frequent classification is based on the type of transduction technique used in the sensor. The three biosensors based on transduction are optical, electrochemical, and mass-based biosensors. The three categories are discussed below in detail.

4.4.1 ELECTROCHEMICAL BIOSENSORS

Electrochemical techniques are the most commonly used for making biosensors. A coating of biological molecules onto a probing surface makes electrochemical biosensors to manufacture this biosensor. A noninterfering membrane is used to keep the sensor molecules in place. The sensing molecules then respond to the substance being detected in a way that provides an electrical signal proportionate to the quantity being measured. Potentiometric, amperometric, conductometric, and impedimetric transducers are electrochemical biosensors that convert chemical information into a measurable electrical signal.[50] The main disadvantage is electromagnetic interference due to nearby electromagnetic sources.

4.4.2 OPTICAL BIOSENSORS

Another type is the optical biosensor. The four standard methods are SPR; optical fiber; red, green, and blue (RGB); and the RAMAN-based biosensors. The last method is the SPR. SPR is based on the principle of variation of the contact surface's refractive index. The reaction changes the quality by binding antibodies to the metal layer, which changes the medium's refractive index in contact with this layer.[1,48] Different types of biosensors are shown in Figure 4.2.

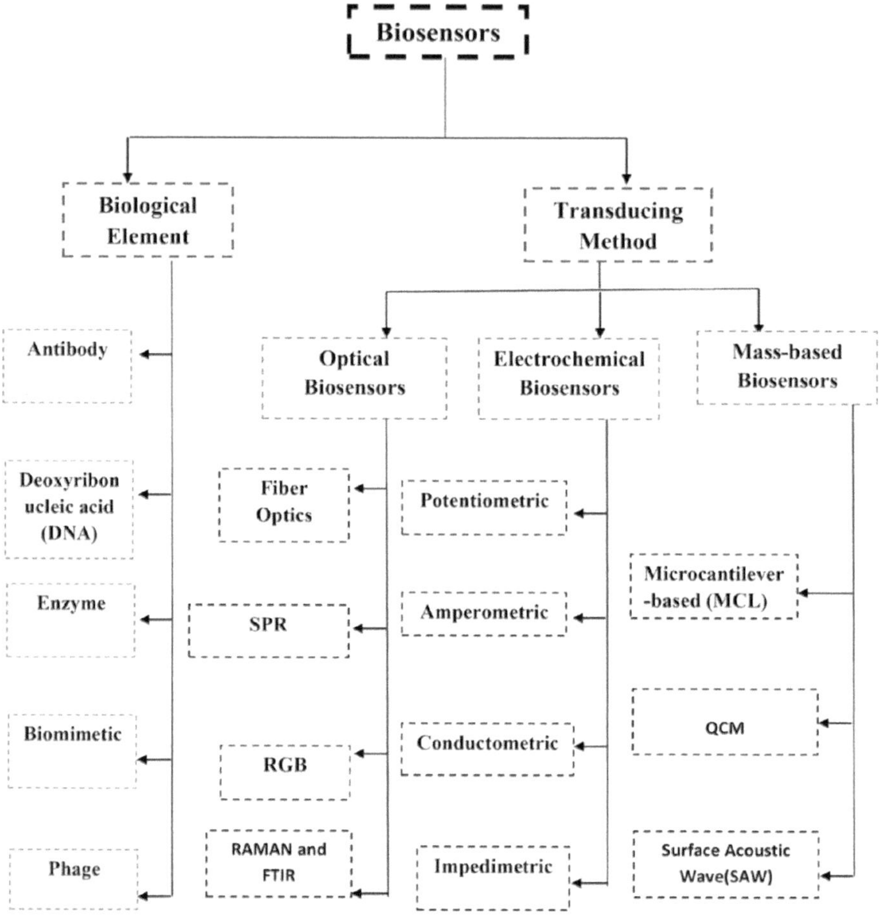

FIGURE 4.2 Different types of biosensors.

The fact that optical biosensors are nonelectrical is one of their key benefits. By altering the wavelength of the light, they can evaluate many components on a single layer.

4.4.3 MASS-BASED BIOSENSORS

Mass-based biosensors are based on a method in which the analyte changes the mass of the biosensor, and it can be measured using different techniques. Some of the methods are SAW, MCL, and QCM, which use the piezoelectric material to make this biosensor based on the principle of sound vibrations. A piezoelectric biosensor transduces the mechanical force signal into an electrical signal. The biological elements are attached to the piezoelectric biosensor's surface. The vibrations of the detecting molecules are converted into proportionate electrical signals by the piezoelectric biosensor. It is a mass-to-frequency converter.[48]

4.5 APPLICATIONS OF IoT-BASED BIOSENSORS

Biosensors have proven essential in medicine, clinical analysis, and general health monitoring since their invention in the early 1950s.[50] Biosensors have several benefits over lab-based technology, including compact size, low cost, rapid findings, and ease of use. Agriculture, pollution management, manufacturing processing, and food dispensation are other application areas of biosensors apart from health and medicine.[50] As a result, the following list of prospective sectors where IoT-based biosensors are recently being utilized and extensively used are discussed below.

4.5.1 IoT-BASED CLINICAL DISEASE APPLICATIONS

Clinical diagnostics is one of the prime applications of biosensors. The IoT helps in sending the data to remote locations via the internet. The analysis of glucose and lactic acid levels in clinics and biochemical labs was mainly done by electrochemical sensors.[30,45] The primary commercial use of biosensors is for the self-monitoring of blood glucose. It has the advantage that the blood does not get polluted or diluted, and precise results are obtained. Previously, point-of-care devices were single-use applications, which meant that the test could only be conducted once and the sensor had to be discarded. However, advancements in this sector provide reusable sensors for enhanced patient care. Some examples of the diagnostic use in IoTB are discussed below.

4.5.1.1 IoTB IN CARDIOVASCULAR DISEASES

The bulk of the published studies on cardiovascular illnesses is about biosensors for measuring cholesterol as sensing elements for the construction of a cholesterol biosensor to measure free cholesterol and total cholesterol, primarily cholesterol esterase and cholesterol oxidase.[2] Electrochemical probes have been used successfully to assess system cholesterol levels.[5,14] Many optical sensors have been utilized to sense cholesterol, comprising monitoring, luminescence, dye color change, fluorescence, and others, depending on the number and accuracy of optical approaches.[2] Other biomarkers for cardiovascular disease are also tested. CRP detection is based mainly on immunosensing methods that include electrochemical, optical, and acoustic transducers and methods for simultaneous analyte measurement.[6,33] Streptavidin polystyrene microspheres were integrated onto the electrode surface of SPEs to improve the analytical response of virus nanoparticles, and cardiac troponin T was utilized for troponin I's very high sensitivity and selectivity diagnostic, a protein marker for increased risk.[13] Research is also going on in this field to develop the IoTB for cholesterol detection for cardiovascular disease.

4.5.1.2 IoT-BASED BIOSENSORS FOR CANCER APPLICATIONS

Cancer is the main reason for death in industrialized and developing countries worldwide. Smoking is one of the significant cancer-causing behaviors. Breast cancer is the leading cause of mortality among the female population, and lung cancer is the leading cause of mortality among the male population. In economically developed countries, breast cancer is the leading cause of mortality. Solid malignancies are a prime source of disease and mortality because they are based on the useless scientific identification and remedy of metastatic contamination in distant locales.[17,19] Environmental as well as hereditary reasons are also responsible for cancer. Other factors, such as radiation, carcinogenic chemical substances, bacterial or viral infections, and pollutants, are also responsible for cancer.[18]

Bioaffinity-based electrochemical biosensors are primarily used to identify gene mutations of biomarkers and protein biomarkers in cancer biomarker evaluation. The amperometric, potentiometric, and impedimetric/conductivity gadgets are based on antibodies. These provide high selectivity and sensitivity for early cancer diagnosis and are an example of electrochemical affinity sensors. Amperometric and potentiometric transducers were most commonly used in earlier times, but impedance-based transducers have gained much interest in recent years. The reason is that they are label-free detecting sensors. These sensors are in their early development stages.[28] Apart from antibodies, the electrochemical gadgets based on DNA hybridization were created and hired for most cancer gene mutation detection. An unmarried-stranded DNA series is fixed at the electrode surface. Here, the DNA hybridization takes place on this tool.[29] Banerjee et al. have shown the use of IoT in cancer detection.[10]

4.5.1.3 IoTB FOR DIABETIC DISEASE

Invasive and noninvasive glucose monitoring devices are available. The first known biosensor[3] was a glucose biosensor, and since then, many other glucose biosensors, including implanted sensors for monitoring glucose in blood or tissue, have been developed. Small, noninvasive glucose sensors are now available to monitor interstitial glucose concentrations in subcutaneous fat.[11] Device miniaturization, elimination of oxygen dependency, long-term stability, user convenience, and biocompatibility are essential for glucose monitoring sensors. Sensor responsiveness is reduced as small molecules diffuse from the sample via the polyurethane sensor transmembrane domain. To tackle the problem, glucose biosensors have been employed with microdialysis or ultrafiltration technologies. Glucose oxidase-based electrochemical technologies are used in commercially marketed invasive glucose monitors, which have electrochemical sensors implanted into the glomerular capillary region. The most widely used electrochemical sensor is the glucose biomarker. It is built on a screen-printed amperometric temporary electrode. This type of biosensor is commonly used for personal glucose testing worldwide. They were bringing onsite analysis to diagnosis. Systems based on IoT have been recently introduced to improve the quality of healthcare services. The number of developed IoT-based blood glucose monitoring systems, on the other hand, is limited, and the present systems have significant disadvantages. Gia et al. have investigated the possibility of an invasive and continuous glucose monitoring system using an IoT-based technique in their work. They have created an IoT-based system architecture, from a sensing element to a backend system, to display real-time glucose, body temperature, and contextual data to end-users such as diabetes healthcare providers and patients in graphical and human-readable form.[23]

4.5.2 IoT-BASED ENVIRONMENTAL MONITORING

Monitoring environmental contamination is one of the most typical packages of biosensors. Because of the close link between environmental pollution and human health/socioeconomic development, the monitoring system has been one of the European and worldwide concerns. Biosensors have been widely used in this sector as cost-effective, rapid, in-situ, and real-time analytical procedures. The current development of biosensors using novel transduction materials generated from nanotechnology and for multiplexed contaminant detection, employing interdisciplinary expertise, emphasizes the need for portable, quick, and smart biosensing devices.

Although environmental biosensors may be built utilizing enhanced nanomaterials and innovative nanocomposites, more emphasis has been placed on in-situ and legitimate monitoring of contaminants using alternative technologies. As disclosed in some daily news, environmental monitoring has recently been a field of interest again for the application of drones, primarily in tracking water and air quality, agricultural surveillance, and volcano gas measured values. A few scientific papers seem to be consistent with the current literature. Considering the biosensors mentioned in this review article, we can infer that electrical and enzyme biosensors are primarily utilized for pollution management, based on their highlighted analytical properties. The primary constraint discovered for current environmental biosensors is a lack of application in genuine environmental samples since the bulk of reported "environmental biosensors" have been applied to tap water or synthetic samples.[44] As a result, there are few commercialized biosensors for pollution management compared to clinical applications, which should be due to the multidisciplinary context of manufacture and constraints on in-situ operation and analytical performance, particularly repeatability. They have implemented the IoTB to solve the constraints discovered.[31]

4.5.3 IoT-BASED FOOD AND MICROBIOLOGY INDUSTRIES

Food safety is a significant health concern in both animal and human lives, and it is one of the primary goals of food analysis. The advancement of analytical techniques in food safety assures that it flourishes in response to the growing interest in and attention to food supply safety problems. However, traditional techniques of food safety analysis are labor-intensive, time-consuming, and require the use of expert professionals. Microfluidics' application in food safety analysis throws fresh light on the practical and speedy detection of foodborne toxins, allergies, pathogens, toxic compounds, heavy metals, and other pollutants.[35]

Immunoanalysis of bacterial infections and communication of results via a mobile device were used to show healthcare coverage via IoT to prevent contamination of food meals. Because wireless internet has grown so widely over the earth, it may efficiently be utilized to disseminate information about human health. Indeed, certain institutes have proven to measure physical and vital signs of the body (for example, temperature, heartbeat, and pulse) for monitoring via a smartphone. However, biological analytes, including proteins and microorganisms, are still difficult to quantify using antigen-antibody binding due to technical constraints. The terminal unit for the detection of infections, for example, is typically large, insensitive, or even costly for advanced instruments. Seo et al. have investigated the usage of a CMOS image sensor (CIS) for recognizing the signal produced by an immunoanalytical device, which might overcome the detection problem preventing IoT monitoring. In their work, a pocket-sized immunesensor device with CIS detection techniques was created and subsequently used to determine a foodborne disease, Vibrio species, in actual samples. The analysis was monitored and managed using a smartphone, sending the results as data to an internet server for public viewing. This is the first study to demonstrate pathogen surveillance using IoT in health to the best of our knowledge.[46]

Biosensors are used in ion exchange retrieval, which detects modifications in organic composition. For instance, a glutamate biosensor was utilized to analyze the ion trade retrieval of an isoelectric glutamate liquor supernatant. The fermentation procedure is a byzantine procedure with several vital variables, the maximum of which can be brutal to evaluate in real-time. Therefore, on-the-fly monitoring of key metabolites is vital for fast optimization and control of biological techniques. Because of their simplicity and speedy reaction, biosensors have become a hobby in the online tracking of fermentation approaches. Different applications of IoTB are shown in Figure 4.3.

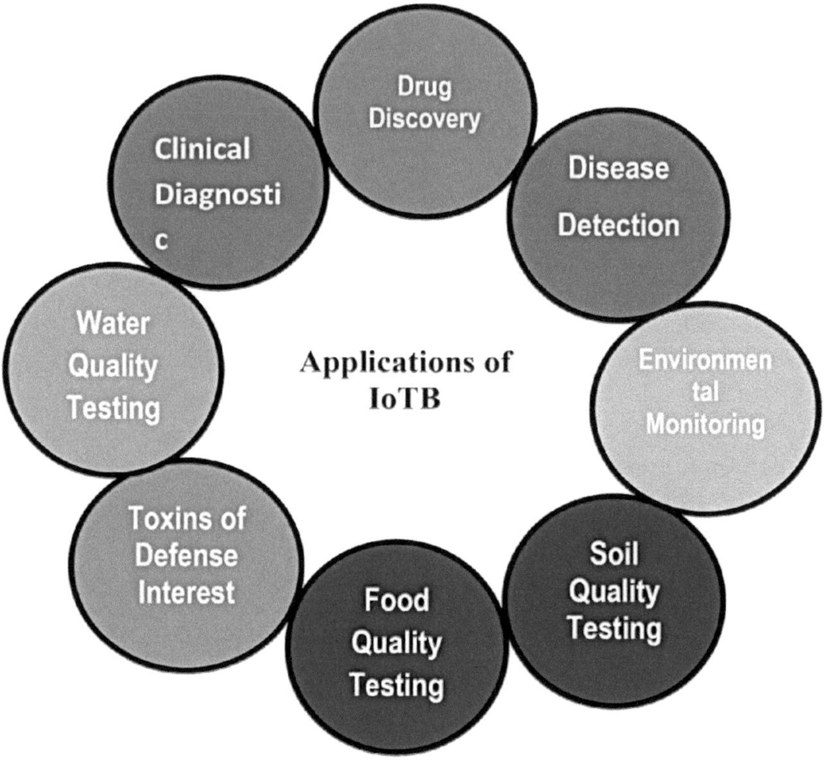

FIGURE 4.3 Different IoTB applications.

Traditional chemical experimentation and spectroscopy techniques have barriers because of human fatigue, which is costly and time-consuming. The food zone might benefit from value-effective options for meal authentication and tracking that consist of objective and constant measurement of food products. As a result, the development of biosensors in reaction to the desire for simple, real-time, selective, and specific value strategies seems to be fantastic. Kuila et al.[33] used enzymatic biosensors primarily based on cobalt phthalocyanine to discover that beer was getting old. The potential of these biosensors to reveal the growing age of beer in the garage has been confirmed. Pathogens in food can be detected using biosensors. A bioindicator of fecal infection in food is the presence of Escherichia coli in greens.[36,37] Using potentiometric alternating biosensing systems, E. coli was detected by detecting pH modifications caused by ammonia (via urease–E.

coli antibody conjugate). The liquid segment is received via washing greens consisting of sliced carrots and lettuce in peptone water. After that, it's amalgamated in a sonicator to split bacterial cells from food objects.[38]

A waft cell became outfitted with a biosensor based on a screen-printed carbon electrode. They have been fixed on electrodes by engulfing enzymes in a photo cross-linkable polymer. An automatic float primarily based biosensor might quantify the three organophosphate pesticides in milk. One of the not unusual meal additives substantially utilized these days is sweetener, which adversely generates undesirable illnesses, including dental cavities, cardiovascular problems, obesity, and type-2 diabetes. It is said that artificial sweeteners are addicting and lure us into consuming extra high-electricity meals unknowingly, thereby promoting weight advantage.[4] The reaction to synthetic sweeteners is heavily reliant on ligand-binding inside the amino-terminal domains of taste receptors. Taste receptor cells have excellent responses to natural and synthetic sweeteners. The flavor epithelium biosensor produced sparse indicators with advantageous waveforms while glucose was administered, whereas sucrose produced nonstop indicators with harmful spikes. Artificial sweeteners elicited more excessive alerts from the taste epithelium, demonstrating that the responses to synthetic sweeteners differed from those to natural sugars in both time and frequency domains.[38]

4.5.4 IoTB IN CLINICAL DIAGNOSIS

Pathogenic microorganisms, including microorganisms, viruses, fungi, and parasites, motivate infectious illnesses. Tuberculosis, meningococcal meningitis, malaria, acquired immunodeficiency syndrome, pneumonia, poliomyelitis, hepatitis, Ebola virus ailment, dengue and chikungunya, American trypanosomiasis (Chagas ailment), leprosy, toxoplasmosis, and leishmaniasis are only a few of the illnesses indexed by the World Health Organization. Although numerous styles of biosensors have been efficaciously created and carried out in the medical career for the prognosis of pathologies along with most cancers,[9] cardiovascular,[8] autoimmune,[9] and neurodegenerative ailments,[21] this overview provides awareness on infectious illnesses. Viruses, including the human papilloma virus[24] (Table 4.2), dengue virus, hepatitis virus,[25] and COVID-19, are infectious carriers which could cause a selection of illnesses in people.[47] Table 4.2 shows different biosensors developed for the detection of various viruses. Bacher et al. have shown the different uses and benefits of the IoTB in virus detection.[7]

TABLE 4.2 Biosensors Used for Virus Detection.

S. No.	Virus type	Biosensor used
1	Human papiloma virus	Electrochemical Genosensor
2	Hepatitis B virus	Electrochemical Genosensor
3	Dengue virus	Electrochemical Immunosensor
4	COVID-19 virus	Electrochemical Sensor

4.5.5 IoTB IN DRUG DISCOVERY

Biosensors are increasingly being used to affirm 'hits' from useful-based assays using a receptor-binding test. Label-loose era builders see their information as having high information costs, which may assist pharmaceutical researchers in making higher judgments all through lead optimization. When combined with quantifiable structural information, these records can have a larger impact (e.g., affinity quantitative structure-activity relationship (QSAR), thermodynamic QSAR, and kinetic QSAR).[20] Calorimetry technology offers sophisticated information and direct thermodynamic measurements for nearly all interactions. It is label- and immobilization-unfastened, and its specific applicability is the most high-quality benefit. Calorimetry might be used in drug development as a high-content material discovery technique and a tool to assess 'hard goals' for which no assays exist. People were skeptical some years ago. The IoTB can play a vital role and help increase the research speed for new drug discovery.[20] Table 4.3 shows the benefit of the IoTB in drug discovery.

TABLE 4.3 Key Application Areas and Benefits of Biosensors in Drug Discovery.

Application	Benefit
Target identification and validation	Receptor deorphaning, binding-based confirmation of pathway-based screen hits
Binding	Hit validation/secondary screening
Lead characterization	Optimization (affinity QSAR and kinetic QSAR), Mode of action
Assay design	QA/QC of assay components used in HTS
Rank order, affinity, kinetics, and specificity	High information content-triaging of real positives
Early ADME	Predictive bioavailability data
Concentration measurement	Improved accuracy and precision in manufacturing and QA/QC for HTS assays

4.5.6 IoTB IN THE AGRICULTURE INDUSTRY

Biosensors in the field of agriculture are generally used to detect pesticides. Agriculture is nothing but the production of crops. Therefore, it is essential to maintain the quality of crops. During the production of crops, various pesticides are used to control pests and disease-causing microbes. Biosensors can also be used in the agriculture field to maintain the quality of crops. Various biosensors like microbial, electrochemical, optical, and others protect the crops. A biosensor has been developed to detect fungi causing damage to crops. A nanosensor application in agriculture is a developed technique. Nanosensor is a small device that can sort out problems regarding the wastage of crops in agriculture. They are used to detect the presence of pathogens, microorganisms, weeds, various gases, and others. It keeps the food maintained throughout the process.[8]

Hence, a finished product of good quality can be reached to the customers. A nanosensor detects physical, chemical, physiological, and biological signals. The waste produced in agriculture during and after harvesting is due to improper practices. The use of nanosensors can reduce this waste. Nanosensors are one of the emerging and new technologies used in agriculture. This is also useful in detecting environmental conditions. Environmental changes are essential factors which affect the quality of the crops. Biosensors detect sudden changes in the environment, and hence, the possible loss of crops can be reduced to some extent. Nanosensors, nanomaterials, nanoparticles, and many other emerging techniques have a massive scope in agriculture.[13] The photosensitive impact and the Internet of Things (IoT) have developed a novel, nondestructive approach for assessing pesticide poisoning in fruits and vegetables. Consequently, customers can have faith in the safety of the product. Safe greens are a portable, practical, accurate, and user-friendly answer to this problem. It is a simple gadget that comes with a mobile phone. The app was developed to aid users in selecting the best items, such as high-quality fruits and vegetables. This is a fresh perspective on IoTB to detect pesticides in fruits and vegetables. This could be a helpful tool in improving the public's general health.

4.5.7 IoTB IN WATER TESTING

Monitoring water in distribution networks is difficult when managing widespread wireless sensor networks (WSNs). A water supply network for detecting chlorine content was presented. Solar-enabled distributed

WSNs have been proposed for protective measures such as pH, turbidity, and oxygen density. Water levels at various locations are monitored using solar-powered sensor nodes and a base station. The approach presented in the research has the advantages of flexibility, low carbon emissions, and low power usage. They offer a combined system for assessing water and air quality that includes additional sensors for detecting air humidity and temperature.

After a comprehensive experimental examination, the United States Environmental Protection Agency found that chemical and biological pollutants used influenced several water parameters evaluated, including turbidity, electrical conductivity, oxidation-reduction potential, and pH. Thus, water quality may be inferred by monitoring and identifying changes in water parameters.[22]

4.5.8 IoTB IN TOXINS OF DEFENSE INTEREST

A vast range of synthetic chemicals, poisons of plant and animal origin, and biological materials—including many disease microorganisms and certain bacterial exotoxins—have been employed or are thought to have the potential to be utilized as warfare weapons. Despite massive efforts to build biosensors, commercially available devices can only assess a small number of analytes, particularly hazardous chemicals. The inherent destabilization and availability of appropriate proteins and the essentially irrevocable nature of the proteins encoded for receptors and antibodies are factors that currently mitigate against the use of enzymes, organic receptors, or antibody-based biosensors for unmonitored ongoing environmental observation of toxic materials. In addition, when operating in a hazardous environment, assays employing antibodies or DNA-based biosensors take time. On the other hand, biosensors may be utilized for exceptionally high sensitivity and onsite specificity detection of contaminated water by specific harmful chemicals.[40]

The use of microbes, viruses, other types of organisms, or poisons produced from live things as a bioweapon to inflict disease or sickness on a human can be described as bioterrorism. Because the consequences of a biological weapon are far more unpredictable, it may eventually become more lethal than a traditional chemical weapon. A variety of harmful microorganisms have been identified as potential biological warfare weapons.

There are hazardous bacteria in Francisella tularensis, Clostridium botulinum, Salmonella typhimurium, and Yersinia pestis. Many bioagents, like Ebola, Venezuelan equine encephalitis, influenza, and Marburg viruses, are of minor importance, even though these viruses are severe and the death rate is relatively high. However, their role on the biological warfare agents (BWAs) list is lower due to the difficulty in their preparation. Doctors observed that civilians died of anthrax pneumonia in 1979, following the accident implicating B. anthracis in the Soviet Union. Anthrax pneumonia generally kills quickly. Bacilli reproduce quickly and create a poison that prevents respiration.While antibiotics are effective against this bacillus, supplies sufficient to meet the treatment needs of a large metropolitan population following an attack would need to be delivered and distributed within 24 to 48 hours of infection. The National Pharmaceutical Stockpile Program is intended to facilitate such an emergency reaction. BWA is less expensive to produce than chemical agents, and a terrorist or military strike utilizing BWA is more successful in terms of both the dangerous region's range and the number of expected victims. The infectious dosage (ID) (quantity of organism required for infection breakout) varies by the agent. Aerosol (particles 1–10 m) inhalation is often capable of causing illness with a smaller ID for the given BWA. A biosensor is a reagent-free analytical instrument comprised of a biorecognition molecule immobilized on a signal transducer. The biorecognition molecule, an enzyme, antibody, DNA sequence, peptide, or even a bacterium, allows the biosensor to be highly selective, allowing it to pick out the desired molecule from a grid of many others. The signal transducer calculates the magnitude of the biorecognition event and turns it into an electrical signal that an alarm can output to an end-user.[34] The IoTB can be very helpful in this field as one has to deal with toxic chemicals, and with the help of the IoTB, the toxins can be measured safely and quickly.

4.6 BENEFITS OF IoT IN THE BIOSENSOR FIELD

a. **Low-cost technology:** The rapid rise of the IoT and accompanying technologies has put downward pressure on the associated hardware and software. The development of successful low-cost technology-enabled care solutions is dependent on a variety of aspects that are either related to IoT sensors, which must have: (a) low power consumption to extend battery life; (b) unobtrusive physical characteristics; (c) robustness to minimize maintenance; (d) wireless connectivity to facilitate networking using widely accepted standards; (e) data preprocessing capabilities to reduce computational load on gateways and the cloud and to the network itself, which must support the deployment and management of a large number of low-cost sensors to enable better care at a lower cost.[41]

b. **Modular, interoperable, and expandable solution:** Interconnection and networking of a large number of heterogeneous smart objects and IoT solutions that use various wireless communication technologies (Bluetooth, Radio Frequency ID, Zigbee, 802.11, 802.15.4, and others) run a variety of often proprietary protocols and applications and have few exposed interfaces that can be required for practical and usable IoT applications. Many polls show that vendor lock-in and complex security and manageability processes stymie the widespread adoption of IoT technology. Various IoT platforms, either domain-specialized (e.g., UniversAAL3) or general-purpose, are developing to address interoperability difficulties (e.g., FI-WARE4 and GoogleApp Engine 5). As a result, developers rely on current platforms and services, and new platforms must be compatible with existing ones. Furthermore, many standardized "IoT communication protocols" have been developed to address interoperability and fragmentation difficulties. The M.Q. Telemetry Transport protocol is one such machine-to-machine (M2M) communication protocol that has recently been defined by OASIS (also standardized as ISO/IEC 20922) and is already being used in a variety of sectors such as eHealth. The OASIS standard DPWS facilitates interaction with resource-constrained devices and has been intensively researched in various fields, including eHealth and smart homes. CoAP,[52] an IETF standard, provides an alternative web transport protocol that permits the integration of limited IoT nodes via lightweight interactions.[41]

c. **Big data and machine learning:** Smart healthcare platforms can continually monitor the health state and activities of patients and the elderly and the safety and security of their surroundings using ambient sensors and wearable devices. This continual flow of data, together with people's medical histories, may provide tailored diagnoses and help automate crucial chores like medical data archiving and evaluating the success of medical therapies. Recent advancements in artificial intelligence and extensive data processing technology have enabled dependable, accurate, and resilient data recording and processing infrastructures. The requirements of the decision-making process, which describe the data to be gathered and evidence to be retrieved to assist intelligent decision-making, can guide the analysis of such data. The following prerequisites must be met to support this: In modular and interoperable data ingestion, (i) devices from various manufacturers contribute to a standard data model; (ii) parallelization and data stream processing in all steps from data acquisition to storage and processing; (iii) use of data analytics and artificial intelligence to support automatic monitoring and decision-making; and (iv) privacy-aware data processing to ensure consistent data encryption, database security, and secure communication channels.[41]

d. **Security and privacy:** The complex interface schemes that arise at some point of the regular, ordinary use of smart healthcare answers (e.g., sufferers/users with their caregivers/clinical specialists and others), in addition to the private/sensitive nature of the handled facts, are two considerable troubles regarding the security and privacy of relevant programs. The safety of the framework's M2M and machine-to-human interactions demands the incorporation of sturdy security and privacy protections, including seamless authentication and authorization services. IoT devices were developed with affordable value, low electricity use, and simplicity of setup, use, and interconnection, rather than safety. However, because fitness tracking structures and smart houses and workplaces may also contain valuable belongings, guarding them in opposition to vicious assaults is critical. The following is a list of safety and privacy dangers to be aware of:[41]

    a. Small, low-fee, network-connected devices have inadequate safety functions.
    b. Because of the community's restrained processing abilities, computationally steeply-priced real-time operations (along with circumstance reasoning) must be relocated to the cloud.

c. Secure verbal exchange within the clever domestic employs various protocols (Wireless Fidelity (WiFi), Bluetooth, ZigBee, and others), which can cause an expansion of assaults.
d. Smart domestic sensors, for example, create a tremendous amount of personal records and metadata.
e. Consent is needed for the safe transfer of anonymized information.
f. A smart home's physical safety is connected to protecting sensitive structures for the citizens' healthcare.
g. Communication with the backend processing systems ought to be costly.
h. Interactions with third parties such as clinics and hospitals require that their records and metadata must be cozy and privacy-retaining.
i. Precise accounting and authorization mechanisms must be suited to all actors (from the elderly and scientific experts to sensors and backend systems).

4.7 TECHNOLOGY ADVANCEMENTS AND CHALLENGES IN THE AREA OF IoT BIOSENSING

Calorimetry technologies offer sophisticated records and direct thermodynamic measurements for nearly all capability interactions. It is label and immobilization-free, and its worldwide applicability is the maximum top-notch advantage. Calorimetry can be used in drug improvement as a high-content material discovery approach and a tool to evaluate 'tough objectives' for which no assays exist. People had been skeptical of electrophysiology at high-throughput quotes a few years ago, and Vivactis is now effectively overcoming the identical reservations about calorimetry.[26] If those troubles are resolved, electrochemical strategies could become more widely used. Low detection limits are essential to examine full-size biomolecules, which can be advanced by using nanomaterials consisting of nanoporous gold, graphene, or CNTs. The pore length of nanoporous gold may be customized between five and seven hundred nm, which is vast enough to hold enormous biomolecules. Nanoporous gold (NPG) is an attractive material for biosensing due to its excessive floor vicinity, excellent conductivity, and biocompatibility.[27]

NPG's usefulness to research programs is presently limited due to the intricacy of the manufacturing process. Alternatively, carbon nanomaterials can be synthesized at a relatively modest cost. Compared to carbon substances like CNTs or structured mesoporous carbon, graphene has a higher sensitivity for many biomolecules. In addition, graphene's specific features (excessive conductivity, excessive surface area, excellent mechanical features, ease of functionalization, and scalability) and occasional manufacturing fee make it an attractive cloth for biosensor manufacturing. Wearable sensors and personalized health ought to gain from graphene in noninvasive biosensors. Such devices present boundaries such as accelerated consolation and analytical performance.[15]

Additive manufacturing can also build geometrically complex gadgets in a digitally managed process. A variety of structures and electrodes have been manufactured using additive manufacturing methods. However, additive production remains in its early stages, and more thorough research is needed. For example, the immobilization of enzymes deep inside pores and channels may additionally motivate substrate delivery problems. The use of 3D-revealed bespoke microfluidic gadgets may be able to steer clear of those transportation constraints. On the other hand, the electrochemical performance of 3D-printed electrodes has been inferior compared to electrodes made using different traditional methods. Furthermore, the high price of consumables and equipment required to fabricate 3D-revealed metal electrodes and the difficulties of manufacturing porous systems may additionally restrict biosensing packages. Because of their low price, ease of fabrication, and appropriateness for huge-scale manufacturing, 3D-revealed carbon electrodes maintain a giant ability. However, only a tiny variety of biosensors based on 3D-published electrodes have been posted. More studies are needed to generate 3D-revealed electrodes on an extensive scale and the performance required for clinically widespread analytes. The different challenges faced in the IoT biosensing field are as follows:

a. **Servicing and Maintenance Cost:** Recently, there have been significant technological breakthroughs that require the ongoing upgradation of IoT-based devices from time to time. Every IoT-based system includes many medical devices and sensors that are linked together. This entails expensive maintenance, repair, and upgrade expenses, which may influence the company and the end-user financials. As a result, sensors that can be operated with a lower maintenance cost must be included.[43]

b. **Power Consumption:** The majority of IoT devices are powered by batteries. Once a sensor is installed, replacing the battery is challenging. As a result, a high-capacity battery was employed to power such a system. However, researchers worldwide are actively attempting to build healthcare gadgets that can produce electricity. Integrating the IoT system with renewable energy sources might be a potential option. To some extent, these systems can assist in alleviating the global energy issue.[43]

c. **Standardization:** Many suppliers manufacture various items in the healthcare business. Most of these products claim to adhere to conventional design standards and processes. There is, however, a lack of validity. As a result, they formed a specialized organization that can standardize these IoTB devices based on communication protocols, data aggregation, and gateway interfaces. The validation and standardization of electronic medical records recorded by IoTB devices must also be thoroughly studied. This can be accomplished by collaborating with researchers from various organizations and standardization bodies such as the Information Technology and Innovation Foundation, the European Telecommunications Standards Institute, the Internet Protocol for Smart Objects, and others to form working groups to standardize the device.[43]

d. **Data Privacy and Security:** The use of cloud computing has changed the concept of real-time monitoring. However, this has increased the vulnerability of healthcare networks to attackers. This may result in the mishandling of patients' valuable information and impact the treatment process. To protect an IoT system from malicious assault, numerous precautions must be taken throughout the design process. Medical and sensor equipment in an IoTB network must analyze and use identity authentication, secure booting, fault tolerance, authorization management, allow listing, password encryption, and secure pairing protocols to avoid an attack. Similarly, network technologies like WiFi, Bluetooth, and Zigbee must be integrated with secure routing algorithms and message integrity verification procedures. Because IoT is a connected network in which each user is linked to the server, any flaw in IoT security services may jeopardize the patient's privacy. This might be remedied by incorporating sophisticated and safe algorithms and cryptographies into a more secure environment.[43]

e. **Environmental Impact:** The creation of an IoTB system necessitates the integration of numerous biological sensors with semiconductor-rich devices. Earth metals and other harmful compounds are commonly used in manufacturing and fabrication. This may hurt the environment. As a result, an appropriate regulatory agency must be established to control and regulate sensor manufacturing. Furthermore, an additional study must be conducted to develop sensors made of biodegradable materials.[43]

f. **Scalability:** Scalability refers to a healthcare device's capacity to respond to changes in the environment. A system with greater scalability operates smoothly and efficiently, utilizing all available resources. As a result, developing a device with more scalability is critical. This increases the efficiency of a system for current and future purposes. An IoTB system is the connectivity of various medical devices, sensors, and actuators that communicate information through the Internet. The absence of homogeneity across an IoTB system's linked devices reduces the system's scalability and must thus be adequately controlled.[43]

g. **Identification:** Healthcare employees need to deal with many sufferers and care for them at the same time. Similarly, if a patient has diverse fitness conditions, they would see many medical doctors. To reduce confusion and keep the unified operation of the healthcare gadget, it was vital to replace the identities of the patient, caregiver, and medical doctor with a single remedy consultation.[43]

h. **Self-Configuration:** IoT devices must provide more power to users by integrating features such as manual setup of the software used. Users could be able to adjust the system settings based on the application's needs and changes in the ambient circumstances.[43]

i. **Continuous Monitoring:** Many healthcare circumstances, such as chronic illnesses, cardiac disorders, and others, necessitate long-term patient monitoring throughout therapy. In such cases, the IoT device must be capable of doing real-time monitoring effectively.[43]

j. **Exploration of New Diseases:** With the short development of the mobile generation, new healthcare applications are being introduced daily. Even though there are many cellular apps for healthcare packages, the sorts of ailments for which those apps were constructed are limited. As a result, there is a need to cover additional ailments that have been disregarded or received low interest in the past. This could broaden the variety of IoTB applications.[43]

4.8 CONCLUSIONS ON BIOSENSING FOR PRESENT AND FUTURE TRENDS

Progress and development in biosensor improvement could ultimately center on nanomaterial generation, which can tackle biocompatibility and biofouling issues. Pesticides, antibiotics, infections, pollutants, proteins, microorganisms, vegetation, animals, meals, soil, air, and water can be detected using smart nanomaterial biosensors online and in real-time. Thus, biosensors are the perfect analytical equipment for pollution tracking, bearing in mind the proper application of statutory policies to guard our biosphere. Biosensor improvement and miniaturization can be at the vanguard of advanced nanomaterial developments and difficulties for many packages. These growing fields could vastly affect the creation of new ultrasensitive biosensing gadgets to cope with future intense pollutant issues that threaten human health and negatively impact other dwelling beings' comforts.

With the advancement of technology in the IoT domain and the incorporation of the point-of-care technology feature in the biosensor field, the requirement for personal care increases, increasing the IoTB. Hence, the different wireless technology advancements used for data communication could give rise to different challenges and facilities for the end-user, a patient. Therefore, the incorporation of the IoT gives many benefits in the field of healthcare status improvement in human society and brings many challenges as technology use increases in the day-to-day life of humans.

KEYWORDS

- **biosensors**
- **biosensing**
- **IoTB (IoT biosensor)**
- **IoT**
- **disease detection**

REFERENCES

1. Abbas, A.; Linman, M. J.; Cheng, Q. New Trends in Instrumental Design for Surface Plasmon Resonance-based Biosensors. *Biosens. Bioelectron* **2011**, *26,* 1815–1824. https://doi.org/10.1016/j.bios.2010.09.030
2. Adami, A.; Mortari, A.; Morganti, E.; Lorenzelli, L. Microfluidic Sample Preparation Methods for the Analysis of Milk Contaminants. *J. Sensors* **2016**, *2016,* 2385267. https://doi.org/10.1155/2016/2385267
3. Algar, W. R.; Tavares, A. J.; Krull, U. J. Beyond Labels: A Review of the Application of Quantum Dots as Integrated Components of Assays, Bioprobes, and Biosensors Utilizing Optical Transduction. *Anal. Chim. Acta* **2010**, *673,* 1–25. https://doi.org/10.1016/j.aca.2010.05.026
4. Alloush, H. M.; Lewis, R. J.; Salisbury, V. C. Bacterial Bioluminescent Biosensors: Applications in Food and Environmental Monitoring. *Anal. Lett.* **2006**, *39,* 1517–1526. https://doi.org/10.1080/00032710600713172
5. Arduini, F.; Cinti, S.; Scognamiglio, V.; Moscone, D. Nanomaterials in Electrochemical Biosensors for Pesticide Detection: Advances and Challenges in Food Analysis. *Microchim. Acta* **2016**, *183,* 2063–2083. https://doi.org/10.1007/s00604-016-1858-8
6. Arya, S. K.; Datta, M.; Malhotra, B. D. Recent Advances in Cholesterol Biosensor. *Biosens. Bioelectron.* **2008,** *23,* 1083–1100. https://doi.org/10.1016/j.bios.2007.10.018
7. Bacher, G.; Bhand, S.; Deshpande, S. Chapter 14—Electrical Biosensors for Virus Detection, In *Advanced Biosensors for Virus Detection*; Khan, R., Parihar, A., Kaushik, A., Kumar, Eds.; Academic Press, 2022; pp 241–259. https://doi.org/10.1016/B978-0-12-824494-4.00004-7
8. Bakhmachuk, A.; Gorbatiuk, O.; Rachkov, A.; Dons'koi, B.; Khristosenko, R.; Ushenin, I.; Peshkova, V.; Soldatkin, A. Surface Plasmon Resonance Investigations of Bioselective Element Based on the Recombinant Protein A for Immunoglobulin Detection. *Nanoscale Res. Lett.* **2017,** *12,* 112. https://doi.org/10.1186/s11671-017-1903-5

9. Balahura, L.-R.; Stefan-Van Staden, R.-I.; Van Staden, J. F.; Aboul-Enein, H. Y. Advances in Immunosensors for Clinical Applications. *J. Immunoassay Immunochem.* **2019,** *40,* 40–51. https://doi.org/10.1080/15321819.2018.1543704
10. Banerjee, A.; Maity, S.; Mastrangelo, C. H. Cancer Detection, IoT, and the Role of Machine Learning in. *Sensors* **2021**.
11. Becker, B.; Cooper, M. A. A Survey of the 2006–2009 Quartz Crystal Microbalance Biosensor Literature. *J. Mol. Recognit.* **2011,** *24,* 754–787. https://doi.org/10.1002/jmr.1117
12. Blum, L.; Marquette, C. Biosensors. In *From the Glucose Electrode to the Biochip BT - Nanoscience: Nanobiotechnology and Nanobiology*; Boisseau, P., Houdy, P., Lahmani, M., Eds.; Springer Berlin Heidelberg: Berlin, Heidelberg, **2009**; pp 871–909. https://doi.org/10.1007/978-3-540-88633-4_16
13. Charpentier, L.; El Murr, N. Amperometric Determination of Cholesterol in Serum with use of a Renewable Surface Peroxidase Electrode. *Anal. Chim. Acta* **1995,** *318,* 89–93. https://doi.org/10.1016/0003-2670(95)00311-8
14. Chen, H.; Jiang, C.; Yu, C.; Zhang, S.; Liu, B.; Kong, J. Protein Chips and Nanomaterials for Application in Tumor Marker Immunoassays. *Biosens. Bioelectron.* **2009,** *24,* 3399–3411. https://doi.org/10.1016/j.bios.2009.03.020
15. Choudhary, M.; Shukla, S.; Narang, J.; Kumar, V.; Govender, P.; Niv, A.; Hussain, C. M.; Wang, R.; Mangala, B.; Rajendran, S. Switchable Graphene-Based Bioelectronics Interfaces. *Chemosensors* **2020,** *8,* 45. https://doi.org/10.3390/chemosensors8020045
16. Clark, L. C. J.; Lyons, C. Electrode Systems for Continuous Monitoring in Cardiovascular Surgery. *Ann. N. Y. Acad. Sci.* **1962,** *102,* 29–45. https://doi.org/10.1111/j.1749-6632.1962.tb13623.x
17. Collings, A. F.; Caruso, F. Biosensors: Recent Advances. *Reports Prog. Phys.* **1997,** *60,* 1397–1445. https://doi.org/10.1088/0034-4885/60/11/005
18. D'Orazio, P. Biosensors in Clinical Chemistry. *Clin. Chim. Acta.* **2003,** *334,* 41–69. https://doi.org/10.1016/s0009-8981(03)00241-9
19. Ding, L.; Du, D.; Zhang, X.; Ju, H. Trends in Cell-Based Electrochemical Biosensors. *Curr. Med. Chem.* **2008,** *15,* 3160–3170. https://doi.org/10.2174/092986708786848514
20. Coates, E. E.; Edupuganti, S.; Chen, G. L.; Happe, M.; Strom, L.; Alicia Widge, M. B. F.; Cox, J. H.; Gordon, I.; Plummer, S.; Ola, A.; Yamshchikov, G.; Andrews, C.; Curate-Ingram, S.; Morgan, P.; Nagar, S.; Collins, M. H.; Bray, A.; Nguyen, T.; Stein, J.; Case, C. L.; Kaltovich, F.; Wycuff, D.; Liang, C. J.; Carlton, K.; Vazquez, S.; Mascola, J. R.; Ledgerwood, J. E. Safety and Immunogenicity of a Trivalent Virus-Like Particle Vaccine Against Western, Eastern, and Venezuelan Equine Encephalitis Viruses: A Phase 1, Open-Label, Dose-Escalation, Randomised Clinical Trial. *Lancet Infect. Dis.* **2022**. https://doi.org/10.1016/S1473-3099(22)00052-4
21. Gajendragad, M. R.; Kamath, K. N.; Anil, P. Y.; Prabhudas, K.; Natarajan, C. Development and Standardization of a Piezo Electric Immunobiosensor for Foot and Mouth Disease Virus Typing. *Vet. Microbiol.* **2001,** *78,* 319–330. https://doi.org/10.1016/s0378-1135(00)00307-2
22. Geetha, S.; Gouthami, S. Internet of Things Enabled Real Time Water Quality Monitoring System. *Smart Water* **2017,** *2,* 1. https://doi.org/10.1186/s40713-017-0005-y

23. Gia, T. N.; Ali, M.; Dhaou, I. B.; Rahmani, A. M.; Westerlund, T.; Liljeberg, P.; Tenhunen, H. IoT-Based Continuous Glucose Monitoring System: A Feasibility Study. *Procedia Comput. Sci.* **2017,** *109,* 327–334. https://doi.org/10.1016/j.procs.2017.05.359
24. Gong, J.-L.; Gong, F.-C.; Zeng, G.-M.; Shen, G.-L.; Yu, R.-Q. An Amperometric Immunosensor for the Newcastle Disease Antibody Assay. *Anal. Lett.* **2003,** *36,* 287–302. https://doi.org/10.1081/AL-120017691
25. Goyal, R. N.; Gupta, V. K.; Chatterjee, S. Voltammetric Biosensors for the Determination of Paracetamol at Carbon Nanotube Modified Pyrolytic Graphite Electrode. *Sensors Actuators B Chem.* **2010,** *149,* 252–258. https://doi.org/10.1016/j.snb.2010.05.019
26. Grieshaber, D.; MacKenzie, R.; Vörös, J.; Reimhult, E. Electrochemical Biosensors - Sensor Principles and Architectures. *Sensors (Basel)* **2008,** *8,* 1400–1458. https://doi.org/10.3390/s80314000
27. Guan, J.-G.; Miao, Y.-Q.; Zhang, Q.-J. Impedimetric Biosensors. *J. Biosci. Bioeng.* **2004,** *97,* 219–226. https://doi.org/10.1016/S1389-1723(04)70195-4
28. Heyduk, E.; Dummit, B.; Chang, Y.-H.; Heyduk, T. Molecular Pincers: Antibody-Based Homogeneous Protein Sensors. *Anal. Chem.* **2008,** *80* (13), 5152–5159.
29. Huang, L.; Muyldermans, S.; Saerens, D. Nanobodies®: Proficient Tools in Diagnostics. *Expert Rev. Mol. Diagn.* **2010,** *10,* 777–785. https://doi.org/10.1586/erm.10.62
30. Ibraheem, A.; Campbell, R. E. Designs and Applications of Fluorescent Protein-Based Biosensors. *Curr. Opin. Chem. Biol.* **2010,** *14,* 30–36. https://doi.org/10.1016/j.cbpa.2009.09.033
31. Justino, C. I. L.; Duarte, A. C.; Rocha-Santos, T. A. P. Recent Progress in Biosensors for Environmental Monitoring: A Review. *Sensors (Switzerland)* **2017,** *17.* https://doi.org/10.3390/s17122918
32. Kress-Rogers, E. Handbook of Biosensors and Electronic Noses. Taylor and Francis/CRC Press, 1996.
33. Kuila, T.; Bose, S.; Khanra, P.; Mishra, A. K.; Kim, N. H.; Lee, J. H. Recent Advances in Graphene-Based Biosensors. *Biosens. Bioelectron.* **2011,** *26,* 4637–4648. https://doi.org/10.1016/j.bios.2011.05.039
34. Kumar, H.; Rani, R. Development of Biosensors for the Detection of Biological Warfare Agents: Its Issues and Challenges. *Sci. Prog.* **2013,** *96,* 294–308. https://doi.org/10.3184/003685013X13777066241280
35. Liu, G. L.; Rosa-Bauza, Y. T.; Salisbury, C. M.; Craik, C.; Ellman, J. A.; Chen, F. F.; Lee, L. P. Peptide-Nanoparticle Hybrid SERS Probes for Optical Detection of Protease Activity. *J. Nanosci. Nanotechnol.* **2007,** *7,* 2323–2330. https://doi.org/10.1166/jnn.2007.444
36. McBride, J. D.; Cooper, M. A. A High Sensitivity Assay for the Inflammatory Marker C-Reactive Protein Employing Acoustic Biosensing. *J. Nanobiotechnology* **2008,** *6,* 5. https://doi.org/10.1186/1477-3155-6-5
37. Meadows, D. Recent Developments with Biosensing Technology and Applications in the Pharmaceutical Industry. *Adv. Drug Deliv. Rev.* **1996,** *21,* 179–189. https://doi.org/10.1016/S0169-409X(96)00406-1

38. Mehrotra, P. Biosensors and Their Applications – A Review. *J. Oral Biol. Craniofacial Res.* **2016,** *6,* 153–159. https://doi.org/10.1016/j.jobcr.2015.12.002
39. Mongra, A. C. Commercial Biosensors : An outlook. *J. Acad. Indus. Res.* **2012,** *1,* 310–312.
40. Paddle, B. M. Biosensors for Chemical and Biological Agents of Defence Interest. *Biosens. Bioelectron.* **1996,** *11,* 1079–1113. https://doi.org/10.1016/0956-5663(96)82333-5
41. Pateraki, M.; Fysarakis, K.; Sakkalis, V.; Spanoudakis, G.; Varlamis, I.; Maniadakis, M.; Lourakis, M.; Ioannidis, S.; Cummins, N.; Schuller, B.; Loutsetis, E.; Koutsouris, D. Biosensors and Internet of Things in Smart Healthcare Applications: Challenges and Opportunities. In *Wearable and Implantable Medical Devices*; 2020. https://doi.org/10.1016/b978-0-12-815369-7.00002-1
42. Pateraki, M.; Fysarakis, K.; Sakkalis, V.; Spanoudakis, G.; Varlamis, I.; Maniadakis, M.; Lourakis, M.; Ioannidis, S.; Cummins, N.; Schuller, B.; Loutsetis, E.; Koutsouris, D. Biosensors and Internet of Things in Smart Healthcare Applications: Challenges and Opportunities. 2019; pp 25–53. https://doi.org/10.1016/B978-0-12-815369-7.00002-1
43. Pradhan, B.; Bhattacharyya, S.; Pal, K. IoT-Based Applications in Healthcare Devices. *J. Healthc. Eng.* **2021,** *2021.* https://doi.org/10.1155/2021/6632599
44. Salouti, M.; Khadivi Derakhshan, F. Biosensors and Nanobiosensors in Environmental Applications BT—Biogenic Nano-Particles and their Use in Agro-ecosystems. In Ghorbanpour, M., Bhargava, P., Varma, A., Choudhary, D. K., Eds.; Springer Singapore: Singapore, 2020; pp 515–591. https://doi.org/10.1007/978-981-15-2985-6_26
45. Scarano, S.; Mascini, M.; Turner, A. P. F.; Minunni, M. Surface Plasmon Resonance Imaging for Affinity-based Biosensors. *Biosens. Bioelectron.* **2010,** *25,* 957–966. https://doi.org/10.1016/j.bios.2009.08.039
46. Seo, S.-M.; Kim, S.-W.; Jeon, J.-W.; Kim, J.-H.; Kim, H.-S.; Cho, J.-H.; Lee, W.-H.; Paek, S.-H. Food Contamination Monitoring via Internet of Things, Exemplified by using Pocket-sized Immunosensor as Terminal Unit. *Sensors Actuators B Chem.* **2016,** 233. https://doi.org/10.1016/j.snb.2016.04.061
47. Suleman, S.; Shukla, S. K.; Malhotra, N.; Bukkitgar, S. D.; Shetti, N. P.; Pilloton, R.; Narang, J.; Nee Tan, Y.; Aminabhavi, T. M. Point of Care Detection of COVID-19: Advancement in Biosensing and Diagnostic Methods. *Chem. Eng. J.* **2021,** *414,* 128759. https://doi.org/10.1016/j.cej.2021.128759
48. Teles, F. R. R.; Fonseca, L. P. Applications of Polymers for Biomolecule Immobilization in Electrochemical Biosensors. *Mater. Sci. Eng. C* **2008,** *28,* 1530–1543. https://doi.org/10.1016/j.msec.2008.04.010
49. Updike, S. J.; Hicks, G. P. The Enzyme Electrode. *Nature* **1967,** *214,* 986–988. https://doi.org/10.1038/214986a0
50. Wijaya, E.; Lenaerts, C.; Maricot, S.; Hastanin, J.; Habraken, S.; Vilcot, J.-P.; Boukherroub, R.; Szunerits, S. Surface Plasmon Resonance-Based Biosensors: From the Development of Different SPR Structures to Novel Surface Functionalization Strategies. *Curr. Opin. Solid State Mater. Sci.* **2011,** *15,* 208–224. https://doi.org/10.1016/j.cossms.2011.05.001

CHAPTER 5

EMERGING USE OF IoT IN MONITORING PESTICIDE POISONING IN THE HUMAN POPULATION

KRITI AHUJA and NISHU GOYAL

Department of Allied Sciences, School of Health Science and Technology, University of Petroleum & Energy Studies, Dehradun, Uttarakhand, India

ABSTRACT

Pesticides are extensively used during agricultural farming for various benefits such as improved productivity, prevention of crop losses due to weeds/insect pests, and vector diseases. However, they pose serious health hazards such as asthma, reduction in lung and thyroid function, infertility, etc. Pesticides may enter the human body through inhalation, swallowing, or absorption through the skin, and cause several health issues such as coughing, difficulty in breathing, and heart ailments. Biomagnification of pesticides in aquatic animals, bottled water, cow milk, and mother milk is now a worldwide concern. Several research papers reviewed pesticide pollution and its impact on human health and highlighted that diagnosis and management of pesticide poisoning is a major challenge. Estimation of pesticides using conventional methods such as gas chromatography–mass spectrometry (GC-MS), liquid chromatography–mass spectrometry (LC-MS), and high-performance liquid chromatography (HPLC) are time consuming and expensive. To address this concern, the use of miniaturized

IoT-Enabled Healthcare Systems: Applications, Benefits, and Challenges
Yatindra Kumar, Shashi Kant Verma, Archana Singh, Krishna Kumar, & Manoj Gupta (Eds.)

devices such as biosensors that offer several advantages such as low cost, rapid detection, ease of use, and portability are examined. The Internet of Things (IoT) comprised of Sensors, Arduino, and liquid crystal display (LCD), is an emerging technology connecting devices in the physical world. It helps to collect and exchange data and transfer over the Internet to solve several real-world problems. In this chapter, we discuss IoT-based devices used for monitoring pesticide levels, detail the implementation challenges of IoT in the farming environment, and explore future opportunities in this area.

5.1 INTRODUCTION

IoT which is referred to as the Internet of Things (IoT) are devices used to exchange data with other devices over the Internet. In a paradigm "everything is taken to be an object" by Grady Booch, IoT is defined as an object-oriented paradigm with the concept that smart objects are able to communicate with each other.[26] These days, IoT is widely used in various sectors throughout the world. One such sector is healthcare sector where IoT is used for monitoring depression and mood, Parkinson's disease, monitoring heart rate, etc.[27] The purpose is to sense the physiological information by sensors and then analyzing it and sending it for suitable decisions. A relatively new area for the use of IoT is pesticide-level monitoring. Pesticides are chemical substances used widely by farmers to kill pests, insects, small animals, and wild plants or weeds to increase crop productivity. Worldwide estimations state that about 1.8 billion people are involved in agriculture.[4] Figure 5.1 shows the top-10 countries for maximum pesticide consumption. Total consumption of pesticides is estimated to be 3 million ton worldwide with China on the top. India stood 13th with the usage of about 52 kton of pesticides with 0.3 kg/ha of the field.[23]

In this chapter, we intend to explore the use and mode of action of IoT-based devices mainly for pesticide detection in human blood and serum samples. The chapter further discusses the implementation challenges of the given technology and explores future opportunities in this area.

5.2 PESTICIDE POISONING IN HUMANS

The pesticide production in India, started in 1952, near Kolkata. India ranks second in terms of pesticide production in Asia after China. However, the

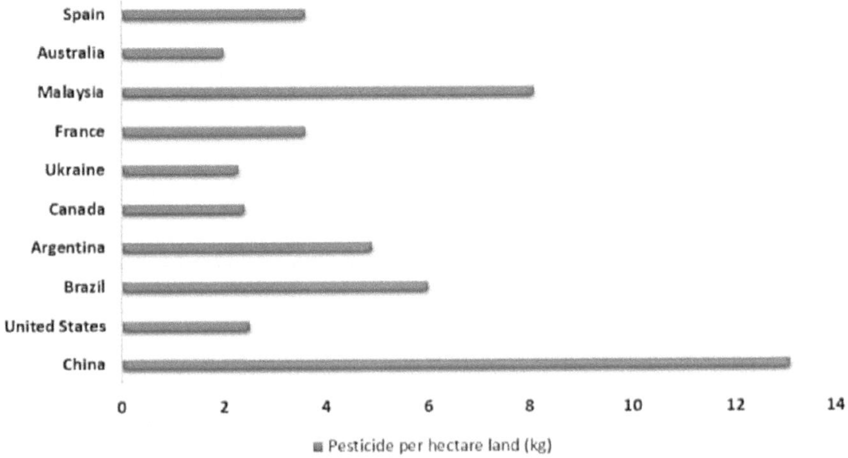

FIGURE 5.1 Top-10 countries' leadings for pesticide usage.

pattern of the usage of pesticides differs in India when compared to the world.[3] These differences can be spotted in the pie charts shown in Figure 5.2. Statistics show that insecticide consumption in India is 1.73 times lower and herbicide consumption is 3 times higher compared to worldwide consumption. Usage of fungicides and others is also higher comparatively.

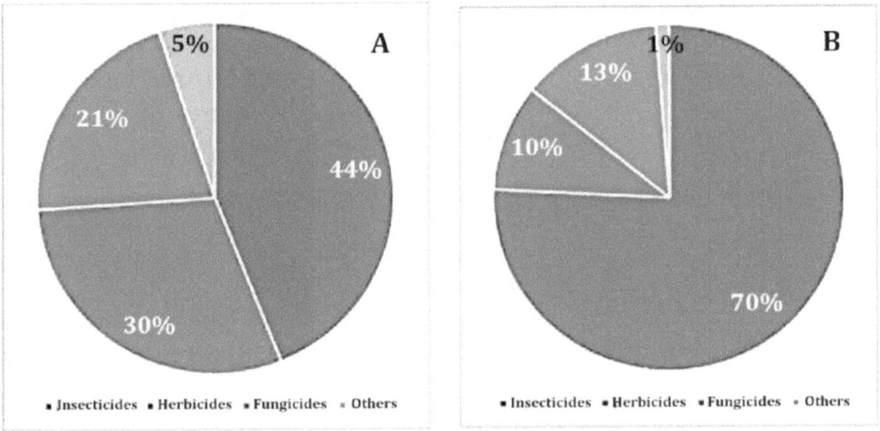

FIGURE 5.2 A comparison on patterns of usage for pesticide types between India and worldwide: (a) pesticide usage in India; (b) pesticide usage worldwide.

Pesticides are marketed and used both nationally and internationally and are reported to have harmful effects both acute and chronic.[19] Acute conditions include rashes, blisters, stinging eyes, diarrhea, nausea, and dizziness whereas chronic disorders include immunotoxicity, neurological and developmental toxicity, cancers, birth defects, reproductive harm, and disruption of the endocrine system.[5] The effects of pesticides vary from one individual to another. The younger generation especially infants are more vulnerable than adults. Hassan et al.,[10] reviewed the effect of pesticide pollution on human health and described various aspects of pesticide usage (Table 5.1). They also discussed old and recent methods for the estimation, detection, and treatment of pesticides. International Agency for Research on Cancer (IARC) broadly works to identify high-risk carcinogens used in the environment and identified specifically two pesticides as carcinogenic. They include arsenical insecticides and 2,3,7,8tetrachlorodibenzo-p-dioxin (TCDD), which is a well-known contaminant of the phenoxy herbicide 2,4,5-trichlorophenoxyacetic acid.[16]

5.3 INTERNET OF THINGS (IoT)

IoT is a novel technology that has completely transformed the perception of the world among people by changing the traditional way of living. IoT has now become such an important part of our lives that it is everywhere and anywhere. When it comes to sensing, processing speed, and storage, IoT uses quantum and nanotechnology.[15] IoT deals with real-life objects or entities such as plants, soil, bodily fluids, clothing, and industrial devices and sends the data to make intelligent analyses. This opens the door to the possibilities of solving various crucial, tedious, and difficult situations in the real world.[2] Figure 5.3 mentions global share for top IoT applications. IoT comprises six layers as follows:

a. ***Perception Layer:*** The perception layer identifies the objects with their associated information. It consists of barcode readers, Radio Frequency Identification (RFID) sensors, and the additional finder arrangement.

b. ***Network Layer:*** The network layer conveys the information collected by the perception layer from the discernment level to the information framework preparing systems.

c. ***Business Layer:*** Its job is to deliver plans of action, stream outlines, government reports, and functional diagrams which in

TABLE 5.1 A Broad Classification of Pesticides and Their Influence on Human Health.

S. No.	Category of the pesticide	Health effects	Commonly used examples
1.	Organochlorine	Hematological and hepatic alterations, and endocrine disorders, affected lipid metabolism and embryonic development. It may also lead to neurodevelopmental effects in children.	Dichlorodiphenyltrichloroethane (DDT), methoxychlor, heptachlor, endosulfan, dieldrin, and dicofol.
2.	Organophosphates (nerve gas)	It attacks the nervous system, especially the brain, affects the function of cholinesterase enzymes. It also leads to the disruption of the normal cellular metabolism of proteins, carbohydrates, and fats. Along with that, it also causes a reduction in insulin secretion.	Glyphosate, dimethoate, parathion, and malathion.
3.	Carbamate	Induction of apoptosis and necrosis in human immune cells, natural killer cells, and also apoptosis in T lymphocytes, and causation of severe cytotoxic and genotoxic effects in hamster ovarian cells.	Aldicarb, carbofuran, and ziram.
4.	Soil Fumigants	Irritation of the skin, eyes, and lungs, and premature birth.	Chlorpicrin, metam sodium, metam potassium, and 1,3-dichloropropene.
5.	Pyrethroids	Headache, fatigue, stinging and itching skin, salivation, involuntary twitching, cancer, tremors, genetic damage, vomiting, and reproductive harm.	Resmethrin, cypermethrin, fenvalerate, and deltamethrin.
6.	Triazines	Endocrine-disrupting effects and reproductive toxicity.	Atrazine, simazine, and ametryn.
7.	Neonicotinoid pesticides	Breast cancer and its effects during developmental periods.	Imidacloprid, thiacloprid, and guadipyr.

turn help in making correct decisions. It covers all of the IoT applications and administrations on board.

d. ***Middleware layer:*** The functionality of the middleware layer is the consistency and administration among various devices. It is comprised of the knowledge and data process frameworks that take the activities that are programmed between the application layer and the hardware layer.

e. ***Application Layer:*** The application layer is the one that deals with real-life applications such as healthcare applications, home services, emergencies, transportation services, education, and many more.

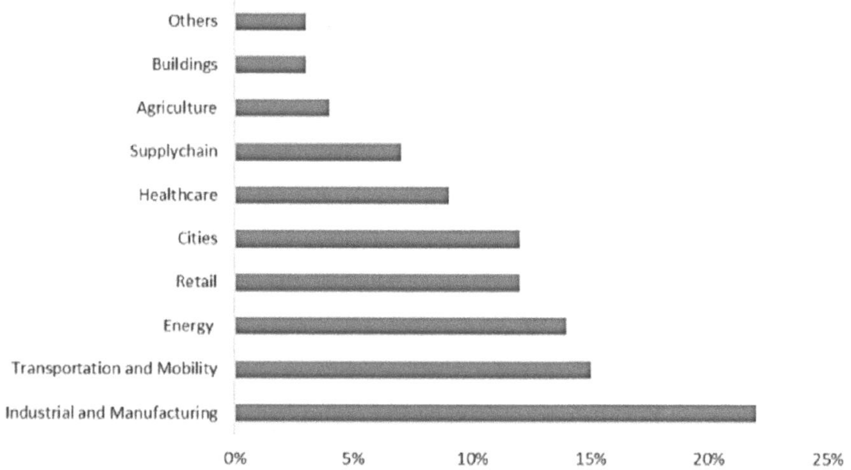

FIGURE 5.3 The top-10 IoT applications are arranged according to their global share in each of them.

IoT hardware has three main components, that is, sensor, Arduino, and liquid crystal display (LCD) as shown in Figure 5.4. The sensors are input-taking devices. They take in the input in the form of data from the environment. The data could be a simple value such as the reading of temperature or a full video. Multiple sensors can be joined together to form a single device, or a device of such kind can have more than just a sensor.

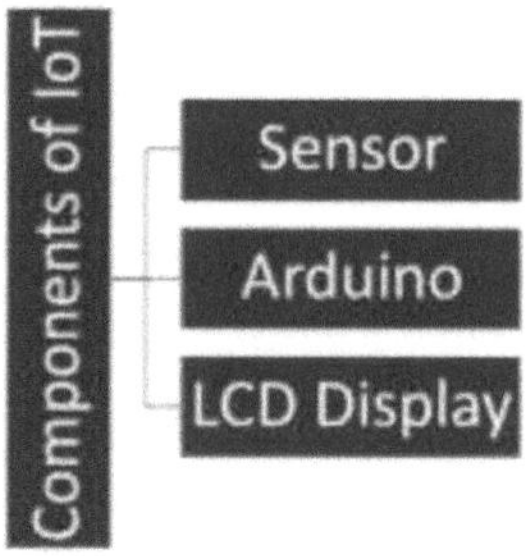

FIGURE 5.4 Components of IoT.

Nonetheless, no matter how complicated the IoT is, the first stage always involves the collection of data using one or more sensors. The open-source microcontroller board is termed the Arduino. The board is composed of various digital and analog input/output (I/O) pins. These could further be connected with different expansion boards and circuits. Some of the features of the Arduino UNO board are (1) 14 digital pins, (2) 6 analog pins, and (3) Arduino Integrated Development Environment (IDE) programmable through a type B cable. The Arduino UNO could be powered by an external 9-V battery or simply a usual USB cable. Arduino UNO is similar to Arduino Leonardo and Nano in the terms of design.[20].

LCD is usually made up of either (A) an active matrix grid or (B) passive matrix grid.[20]

a. Active matrix grid: At each pixel intersection, a transistor is located, needing a reduced amount of electric current to control the luminescence of the pixel. This makes the active-matrix grid way more efficient as the electricity can now easily be switched on and off multiple times. This massively improves the screen refresh time. The active matrix LCD is often also known as a thin film transistor (TFT) display.

b. Passive matrix grid: At each intersection of the grid, pixels are positioned in such a way that they form a grid of conductors. The light of the pixel under consideration is controlled by sending the current across the two conductors.

   The technology of LCD is used for the display of, usually the final results in smaller computers such as the notebook. Unlike the cathode ray technology (CRT), LCDs allow the final display to be comparatively thinner. LCDs work on the principle of blockage of light rather than the emission of it. This makes it more efficient, and less power consuming than the other methods such as light emitting diodes (LEDs) or gas-based displays.[20]

5.4 IoT AND HEALTHCARE

It is often termed HIoT which stands for healthcare IoT. Along with other fields, IoT over time has been providing immense help in the healthcare sector as shown in Figure 5.5. IoT is playing a major role in not only increasing the efficiency of the sector but also making many impossible activities possible. IoT is used in monitoring heart rate, depression, glucose levels, hand hygiene, coagulation, and various others. The devices have other applications such as for cancer patients, connected inhalers for asthma patients, and ingestible sensors for diagnosis of serious diseases. Nearly 150 articles have been published from 2015 to 2020 in the context of IoT and healthcare.[9]

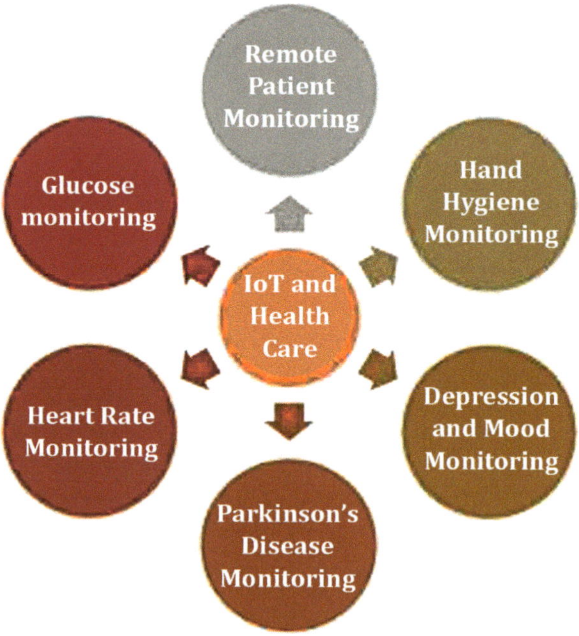

FIGURE 5.5 Various applications of IoT in the healthcare sector.

Some of the examples are as listed in the following:

a. *Remote Patient Monitoring (RPM):* The usage of digital technology usually through IoT portable medical devices to monitor the patient regularly in real time is termed RPM. RPM is of immense use to doctors and patients. It helps the patients to be monitored regularly from the comfort of their homes. The device can detect and inform the caregiver or the doctor in case of a sudden increase or decrease in regular vitals. This gives the provision of help even in emergencies in calculated time.

b. *Hand Hygiene Monitoring:* The other extensively efficient way of preventing diseases is considered to be hand hygiene. World Health Organization (WHO) has specifically stated the guidelines for hand hygiene recommending methods to improve healthcare settings .[30] The IoT devices specifically designed for hand hygiene hence have been a boon to hospitals and other healthcare settings.

c. *Depression and Mood Monitoring:* Close to 500 million people are suffering from mental disorders around the world. The most common being depression, dementia, and schizophrenia. The discovery in this area is however limited. However, mood disorders especially depression are appropriate for the use of IoT technologies .[7]

d. *Parkinson's Disease Monitoring:* Parkinson's disease is a condition in which people are unable to carry out day-to-day activities due to their dead tissues which affects the proper signal transmission. Wearable IoT sensors are being widely used for the monitoring of Parkinson's to perform early and differential diagnoses .[21]

e. *Heart Rate Monitoring:* Most treatments for heart-related problems require monitoring over a longer period .[22] An IoT-based device in such cases acts as a perfect solution. It can measure the heart rate constantly and inform the doctors or the caregivers as and when the situation requires it.

f. *Glucose Monitoring:* As stated by, the Centers for Disease Control and Prevention (CDC) almost 30.3 million residents of the United States of America suffer from diabetes.[8] The IoT devices not only allow regular monitoring of the patients but also warn in emergency and severe cases. Along with this, it is also capable of keeping track of patients' meals and giving suggestions for a healthy diet accordingly.

5.5 IoT AND PESTICIDE DETECTION

Pesticides are being used in agriculture to control diseases, preventing the crop from pests and humans from vector-borne diseases. They are used to eradicate weeds and hence play a major role in preventing them from diseases.[18] However, serious health issues are observed in humans due to the extensive use of these chemicals. Keeping in mind that the food processing industry is the largest in our country and pesticide usage over the safe limit can be harmful, we must focus on solving or addressing this issue.[18] Since IoT-based devices have formed a significant part of the healthcare sector, IoT devices can also be utilized to detect pesticide levels by the use of machine learning and IoT and with the accuracy and performance of these software and hardware stimulations. IoT comprises of sensors such as temperature sensors, gas sensors, humidity, and moisture sensors, and these sensors are linked to the cloud with the help of a network connection. With the use of Wi–Fi modules, this information is further sent to the user.[18] Traditional gas chromatography and other techniques such as liquid chromatography (LC) take more time and are not accurate enough. It also focuses only on the quality. However, we should monitor both quality and quantity. This problem can be resolved with the help of IoT devices. To explain this, we discuss two popular IoT devices as follows.

5.5.1 TWO POPULAR DEVICES—FRESH-O-SNIFF AND THE SMART PEN

i. **Fresh-O-Sniff:** Fresh-O-Sniff is a novel non-destructive device to measure the harmfulness of pesticides in fruits and vegetables. It is a highly efficient, robust, portable, accurate, and extremely user-friendly device that comes with an easy-to-access application for mobiles. This helps the users in selecting premium-quality fruits and vegetables.[13] The principle used by Fresh-O-Sniff is photoconductivity.[13] The optical and electrical phenomenon of photoconductivity is in which the material under consideration becomes more conductive, and electrically because of electromagnetic radiation including gamma radiations, infrared light, ultra-light, or visible light.[1] The photons fall on the device. This leads to the valence band electrons of the semiconductor material being excited to the next band which is the conduction band. The energy of the incident light photons is found to be greater than the bandgap in the semiconductor. This causes the electrons to jump to the conduction band from the valence band. Then the light with enough energy strikes the device. This leads to even more electrons getting excited from the valence band resulting in the formation of charge carriers in very large numbers.[13] The incident wavelength of light may vary in sensitivity. Because of this some of the photocells may not even respond to a particular range of wavelengths. This is why different cells give different response spectral curves.[13] Every IoT-based device is subdivided into hardware and software modules. The hardware module is made up of (1) light-dependent resistor (LDR), (2) microcontroller, and (3) Bluetooth module.[13] Apart from the hardware module, Fresh-O-Sniff comprises a prototype of an LED light source and a glass cuvette. The software module is comprised of the mobile application and the LCD within the mobile application.[13] Here, the illuminated LED light source is made to fall on the glass cuvette or the sample. The LDR sensor measures the refracted light. The results are then sent to cloud storage using the Bluetooth module.[13] The final results are then shown on the mobile application after the analysis. The results tell if the fruit or vegetable under observation is safe to be consumed or not.

ii. **Smart Pen:** Smart pen is extremely easy to use, and is a portable piece of equipment. It makes it comparatively easier for food inspectors to figure out both the quality and quantity of the pesticides present in the fruits and vegetables being consumed. It reduces both time and effort and also gives accurate results for pesticide residues. The device uses Newton's second law of motion as its major principle. Newton's second law of motion states that the rate of change in momentum, p, is directly proportional to the force applied to make changes. It takes place in a straight line. A body having initial velocity u when time $t = 0$, reaching v as velocity when the time reaches t. Then, the change in momentum is mv-mu.[6,29] Hence, momentum is converted from force when the pen is pressed.[29] A smart pen has the following characteristics and components:[29]

   a. Metals Used: Nickel is the metal, which is used for the pen. Nickel is the appropriate choice as it would not react with the fluorescent cyclodextrin.

   b. Light Source: The device uses fluorescent cyclodextrin. This is used as a molecular scanner. The cyclodextrin is capable of producing about 1–50 Hz as a γ-ray. The gamma-cyclodextrin hence has the capability of estimation followed by elimination of pesticide residue from the sample of food under observation. This, therefore, helps in exhausting the mal users.

c. Bluetooth or Digital Meter: The pesticides present can now be found easily using a Bluetooth or digital meter that is connected from a mobile.

The steps followed by the designers in the process of designing the smart pen were as shown in Figure 5.6:

Step 1: The pen is clicked, and the force acts on the fluorescent cyclodextrin. Following, Newton's second law of motion. Newton's second law of motion states that momentum is converted from the force applied. This further leads the light to emit from the pen.

Step 2: Fluorescent cyclodextrin comes into action by being triggered and the light is emitted on the desired fruit or vegetable sample.

Step 3: The next action is molecular scanning. This includes the binding of pesticides with the γ-cyclodextrin used.

Step 4: Here, the amount of the pesticide residue present is detected by the pen through Bluetooth or a digital meter.

Step 5: The final results after the analysis are displayed on the LCD connected to the mobile application. At last, now we can figure out if the fruits and vegetables tested are safe to consume.

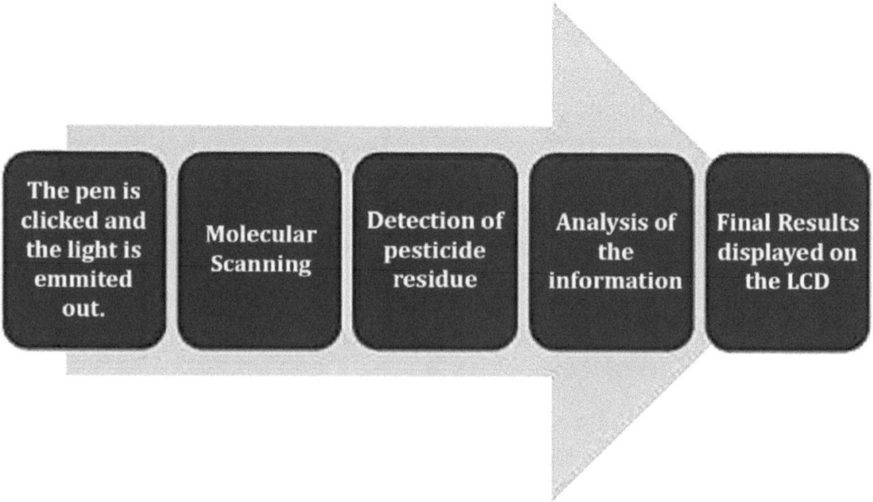

FIGURE 5.6 Stepwise workflow for the usage of smart pen.

IoT when combined with pesticide residue detection services can be used for the detection of pesticides in fruits and vegetables and further research opportunities can be explored in the future. For configuring the light source, detection light sources and sensor/perception layer are used. It detects and changes the spectra of the test strip and converts it to digital data from the detected results. The data is then transmitted to the application layer/network layer by residue detector.[14]

5.6 CURRENT CHALLENGES IN IoT IMPLEMENTATION

IoT has proved to be of great use in various fields and avenues. IoT technology is filling in the gap between the physical and the digital world. This, with its wireless approach, has been the major reason for its success. The strong impact of IoT can be felt in almost all the existing avenues including healthcare systems, agriculture, industries, education, transport, and many more. IoT with artificial Intelligence has brought in surprising expectations with increasing popularity among young and old.

The devices now are becoming more and more intelligent with time. The latest designed models for the implementation of IoT need massive security and privacy, large connectivity, soaring high reliability, and stunted latency. Cisco predicts that almost 500 million devices will be connected to the Internet by the year 2030. The IoT could be said as the most emerging trend in today's world. It is expected that the exponential growth in IoT networking might lead to massive connectivity and greater comfort for us, as humans. However, comfort requires work. With massive connectivity, IoT is supposed to bring up various problems. These current challenges are discussed in detail as follows:

a. **Security and Privacy:** One of the prime challenges is the security and privacy of the data along with the data users. IoT over time has faced serious issues in terms of the privacy of the users and the security of the data. These issues are not only distressing but also fatal at times. Hackers these days use smarter devices and are capable of changing the original instructions given in Arduino. Many technologies these days are focusing on solving the issues from both the hardware as well as software perspectives. The newer protocols for effective communication that include RFID, blockchain, and key management system (KMS) are being developed to improve the privacy, security, and trust in the newer internet technologies such as IoT .[12]

b. **Lack of energy efficiency:** Similar to security and privacy, it is important for IoT devices to be energy efficient in both perspectives, hardware, and software. Researchers and scientists', over time have developed various ways in which the devices could be efficient through the perspective of software. These may include the optimization of link status for communication, using wake–sleep techniques based on network traffic, and reducing the number of hops by developing energy-efficient routing protocols. The fundamental issue of managing power is solved by charging wireless mechanisms .[12]

c. However, when it comes to hardware, there is a profound requirement for highly efficient sensor nodes when it comes to energy so that neither the power efficiency nor the fidelity is compromised.

d. **Scalability:** With the growth in the popularity of IoT, scalability has become a major concern. IoT devices majorly face two main kinds of scalability troubles:

   i. Vertical Scalability: It revolves around the adding and removing of IoT nodes' computing resources.
   ii. Horizontal Scalability: It revolves around the adding or removing of the IoT node itself.

With the proposition of cloud computing and cloud-based architectures, the situations have been controlled to some extent although major challenges remain including IoT nodes expected to provide an increased number of services. These include data storage, functional scalability, fault tolerance, access control, security, and privacy .[12]

i. **Self-organization:** Self-organization in the IoT system talks about the scenario of actively responding during and according to the change in environments with a nicely coordinated and automatic approach to allow it to be within required bounds. These systems automatically react to the change in the environment in conjunction with high-level human policies as well as goals. This could be termed proactive self-management.[12] Nonetheless, there remain many open challenges in this area. Some of the notable research ideas in the future may include cross-platform behavior optimization, large-scale distributed heterogeneous networks of IoT, the development of optimizing self-organizing protocols, and many more.

ii. **Challenges revolving around business:** Because IoT is changing the lives of people, it has been a great area for businessmen to invest in. With this comes certain challenges. One of the major challenges is the increase in demand itself. The manufacturers of IoT are more interested in manufacturing the device as fast as possible than paying attention to other sectors involved, for instance, security. This makes the device prone to hackers and hence riskier.

iii. **Challenges in designing:** The digital devices we use daily, including mobile phones and computers keep getting slimmer. However, the energy used by the battery remains the same. Similarly, the issue lies in the integration and the packaging of tiny chips with lighter weight and lower power consumption.

    Along with that, the designers also need to solve the problems around the design time and the cost modeling of the devices.

iv. **Challenges in the deployment:** The foremost concern while working with the devices and the cloud platform is connectivity. Hence, poor connectivity is another major issue to be dealt with in the future. One of the major areas of IoT becomes data leading to the importance of data management. Hence, along with security and privacy, the developers are also expected to work on how the data is being collected, processed, or stored in an environment.

v. **The Society Challenges:** IoT has been facing major issues when it comes to society. Some of them are listed as follows:

    a. The new devices take time to be brought into use in day-to-day lives.
    b. The demands of the customers are frequently changed. This makes it harder for the industries to cope.
    c. The IoT users are growing with time. Hence, it becomes difficult for IoT to tackle all of them.

d. The underlined issues might become a cause of concern for the buyers leading to a reduction in sales of the IoT products.
e. The integration of new things and older ones requires both time and money.
f. The interface needs to be made simpler. The complex interface could become a barrier for the common man to understand which may again lead to a reduction in sales of the IoT products.

5.7 CONCLUSIONS AND FUTURE PROSPECTS

Modern agriculture has seen a rise in the usage of pesticides. They are effective and economical methods to enhance the yield by improving both the quality and quantity of the growing food. However, the use has been so extensive that it has started to pose a threat to human health.[24] In fact, recent research has depicted pesticides to be a cause of fatal diseases such as cancer due to their exposure to organs such as the brain, lungs, colon, and liver.[25] This makes it extremely important to keep a check on the amount and type of pesticides being used. IoT has the potential to help us perform the detection with limited cost and easy methods. The chapter discussed the role of IoT in pesticide detection and discussed the mechanism of detection. It further explored various applications where IoT-based devices can help us in pesticide detection from blood or serum samples. The scope is wide as the field is still new and there is room for improvement. The work could be done in the correct collection of data by the devices and most importantly in the security and privacy of the IoT devices.

KEYWORDS

- **pesticide**
- **IoT**
- **biomagnification**
- **biosensor**
- **soil**
- **water pollution**

REFERENCES

1. Ahmed, D. M.; Yousef, A. R. M.; Hassan, H. S. A. Relationship Between Electrical Conductivity, Softening and Color of Fuerte Avocado Fruits During Ripening. *Agric. Biol. J. North Am.* 2010. https://doi.org/10.5251/abjna.2010.1.5.878.885
2. Ahmed, S.; Mohammed Ali, A. Overview for Internet of Things: Basics, Components and Applications [Online], n.d. https://www.researchgate.net/publication/332414081
3. Aktar, W.; Sengupta, D.; Chowdhury, A. Impact of Pesticides use in Agriculture: Their Benefits and Hazards. *Interdiscip. Toxicol.* **2009,** *2* (1), 1–12. https://doi.org/10.2478/V10102-009-0001-7
4. Alavanja, M. C. R. Pesticides Use and Exposure Extensive Worldwide. *Rev. Environ. Health* **2009,** *24* (4), 303. https://doi.org/10.1515/REVEH.2009.24.4.303
5. Bonvoisin, T.; Utyasheva, L.; Knipe, D.; Gunnell, D.; Eddleston, M. Suicide by Pesticide Poisoning in India: A Review of Pesticide Regulations and Their Impact on Suicide Trends. *BMC Public Health* **2020,** *20* (1). https://doi.org/10.1186/S12889-020-8339-Z
6. Goren, E.; Galili, I. Newton's Law - A Theory of Motion or Force? *J. Phys.: Conf. Ser.* **2019,** *1287* (1). https://doi.org/10.1088/1742-6596/1287/1/012061
7. Gutierrez, L. J.; Rabbani, K.; Ajayi, O. J.; Gebresilassie, S. K.; Rafferty, J.; Castro, L. A.; Banos, O. Internet of Things for Mental Health: Open Issues in Data Acquisition, Self-Organization, Service Level Agreement, and Identity Management. *Int. J. Environ. Res. Public Health* **2021,** *18* (3), 1–19. https://doi.org/10.3390/ijerph18031327
8. Habibzadeh, H.; Dinesh, K.; Rajabi Shishvan, O.; Boggio-Dandry, A.; Sharma, G.; Soyata, T. A Survey of Healthcare Internet of Things (HIoT): A Clinical Perspective. *IEEE Internet Things J.* **2020,** *7* (1), 53–71. https://doi.org/10.1109/jiot.2019.2946359
9. Haghi Kashani, M.; Madanipour, M.; Nikravan, M.; Asghari, P.; Mahdipour, E. A Systematic Review of IoT in Healthcare: Applications, Techniques, and Trends. *J. Network Comput. Appl.* **2021,** *192,* 103164. https://doi.org/10.1016/J.JNCA.2021.103164
10. Hassaan, M. A.; el Nemr, A. Pesticides Pollution: Classifications, Human Health Impact, Extraction and Treatment Techniques. *Egypt. J. Aquat. Res.* **2020,** *46* (3), 207–220. https://doi.org/10.1016/J.EJAR.2020.08.007
11. *IEEE Xplore Full-Text PDF:* (n.d.). Retrieved May 1, 2022, from https://ieeexplore.ieee.org/stamp/stamp.jsp?arnumber=8972389
12. Imran, M. A.; Zoha, A.; Zhang, L.; Abbasi, Q. H. Grand Challenges in IoT and Sensor Networks. *Front. Commun. Networks* **2020,** *0,* 7. https://doi.org/10.3389/FRCMN.2020.619452
13. Kanmani, R.; Maheswar, R.; Sureshkumar, A.; Mugilan, A.; Praveena, V. In *Non-Destructive Approach to Detect Pesticides in Fruits and Vegetables Using IoT Technology.* 2020 International Conference on Computer Communication and Informatics, ICCCI 2020; 2020. https://doi.org/10.1109/ICCCI48352.2020.9104166
14. Keh, H. C.; Chiang, R. D.; Chang, S. H.; Hung, W. P. Design and Implementation of Pesticide Residues Detection System. *IET Commun.* 2022, https://doi.org/10.1049/cmu2.12355

15. Kumar, S.; Tiwari, P.; Zymbler, M. Internet of Things is a Revolutionary Approach for Future Technology Enhancement: A Review. *J. Big Data* **2019,** *6* (1), 1–21. https://doi.org/10.1186/s40537-019-0268-2/figures/9
16. Leonel, A. C. L. D. S.; Bonan, R. F.; Pinto, M. B. R.; Kowalski, L. P.; Perez, D. E. D. C. The Pesticides use and the Risk for Head and Neck Cancer: A Review of Case-Control Studies. *Med. Oral, Patol. Oral Cirugía Bucal* **2021,** *26* (1), e56. https://doi.org/10.4317/medoral.23962
17. Mira, F.; Alsmadi, I. In *Review of Analysis on IoT Components, Devices and Layers Security, International Conference on Information Science and Communications Technologies: Applications, Trends and Opportunities, ICISCT 2019*, 2019. https://doi.org/10.1109/Icisct47635.2019.9012037.
18. Nagajyothi, D.; Sree, P. H.; Devi, M. R.; Madhavi, G.; Ahammad, S. H. IoT-Based Prediction of Pesticides and Fruits DiseasesDetection Using SVM and Neural Networks. *Int. J. Mech. Eng.* **2022,** *7* (1).
19. Nicolopoulou-Stamati, P.; Maipas, S.; Kotampasi, C.; Stamatis, P.; Hens, L. Chemical Pesticides and Human Health: The Urgent Need for a New Concept in Agriculture. *Front. Public Health* **2016,** *4,* 148. https://doi.org/10.3389/fpubh.2016.00148/bibtex
20. Nikitha, R.; Pavithra, T. K.; Pavithra, V.; Kavyashree, M.; Mrs Vijayalakshmi. IoT Based Solution for Monitoring of Pollution Through Pesticide in Fresh Fruits and Vegetable Available in Market [Online], 2019. http://www.ijser.org
21. Rovini, E.; Maremmani, C.; Cavallo, F. How Wearable Sensors can Support Parkinson's Disease Diagnosis and Treatment: A Systematic Review. *Front. Neurosci.* 2017, *11* (OCT), 555. https://doi.org/10.3389/FNINS.2017.00555/BIBTEX
22. Sahana, S. Khamitkar. . IoT based System for Heart Rate Monitoring. *Int. J. Eng. Res.* **2020,** *V9* (07). https://doi.org/10.17577/IJERTV9IS070673
23. Sharma, A.; Kumar, V.; Shahzad, B.; Tanveer, M.; Sidhu, G. P. S.; Handa, N.; Kohli, S. K.; Yadav, P.; Bali, A. S.; Parihar, R. D.; Dar, O. I.; Singh, K.; Jasrotia, S.; Bakshi, P.; Ramakrishnan, M.; Kumar, S.; Bhardwaj, R.; Thukral, A. K. Worldwide Pesticide Usage and its Impacts on Ecosystem. *SN Appl. Sci.* **2019a,** *1* (11), 1–16. https://doi.org/10.1007/S42452-019-1485-1/TABLES/4
24. Sharma, A.; Kumar, V.; Shahzad, B.; Tanveer, M.; Sidhu, G. P. S.; Handa, N.; Kohli, S. K.; Yadav, P.; Bali, A. S.; Parihar, R. D.; Dar, O. I.; Singh, K.; Jasrotia, S.; Bakshi, P.; Ramakrishnan, M.; Kumar, S.; Bhardwaj, R.; Thukral, A. K. Worldwide Pesticide Usage and its Impacts on Ecosystem. *SN Appl. Sci.* **2019b,** *1* (11). https://doi.org/10.1007/S42452-019-1485-1
25. Sharma, A.; Shukla, A.; Attri, K.; Kumar, M.; Kumar, P.; Suttee, A.; Singh, G.; Barnwal, R. P.; Singla, N. Global Trends in Pesticides: A Looming Threat and Viable Alternatives. *Ecotoxicol. Environ. Safety* **2020,** *201,* 110812. https://doi.org/10.1016/J.ECOENV.2020.110812
26. Sobhan Babu, B.; Srikanth, K.; Ramanjaneyulu, T.; Lakshmi Narayana, I. IoT for Healthcare. *Int. J. Sci. Res.* **2016a**, *5*. https://www.researchgate.net/publication/348929916
27. Sobhan Babu, B.; Srikanth, K.; Ramanjaneyulu, T.; Lakshmi Narayana, I. IoT for Healthcare. *Article Int. J. Sci. Res.* **2016b,** *5*. https://www.researchgate.net/publication/348929916

28. Velu, P.; Palanivelu, V. R.; Apdhulkathar, A. *The Food Processing Industry in India: Challenges and Opportunities*, 2016. https://www.researchgate.net/publication/344403082
29. *View of Detection and Estimation of Pesticides Residue Using a Smart Pen*. (n.d.). Retrieved May 2, 2022, from https://www.journals.grdpublications.com/index.php/ijprse/article/view/38/36
30. Wang, C.; Jiang, W.; Yang, K.; Yu, D.; Newn, J.; Sarsenbayeva, Z.; Goncalves, J.; Kostakos, V. Electronic Monitoring Systems for Hand Hygiene: Systematic Review of Technology. *J. Med. Internet Res.* **2021,** *23* (11). https://doi.org/10.2196/27880

CHAPTER 6

HEALTHCARE SYSTEMS IN INDIA

NARENDRA KUMAR[1], KRISHNA KUMAR[2], and INDRADEO RAM[3]

[1]*School of Computing, DIT University Dehradun, Dehradun, Uttarakhand, India*

[2]*Department of Hydro and Renewable Energy, Indian Institute of Technology, Roorkee, India*

[3]*Department of Computer Science and Engineering, BIT Mesra, Ranchi, Jharkhand, India*

ABSTRACT

Although India has an extensive healthcare system, there are still significant quality gaps between the country's rural and urban regions as well as between the country's public and private healthcare sectors. Despite this, India is becoming an increasingly attractive destination for medical tourists due to the comparatively cheap rates and good quality of its private institutions. Students from other countries who study in India should be prepared to seek sophisticated medical treatment in the country's many private institutions. Because studying in India presents a variety of health issues that students from developed nations may not be used to, it is essential to be familiar with the structure and operation of the Indian healthcare system if you need its assistance. The medical system in India is extensive and, at times, may resemble the rest of the nation in that it is fraught with ambiguity and contradictions. This chapter explains the framework of the Indian healthcare system, including its roles, as well as the techniques that are used to finance healthcare. This chapter explains the many avenues

IoT-Enabled Healthcare Systems: Applications, Benefits, and Challenges
Yatindra Kumar, Shashi Kant Verma, Archana Singh, Krishna Kumar, & Manoj Gupta (Eds.)

and sources of healthcare finance, specifically pertaining to the various modes of healthcare supply. Patients are responsible for paying the bulk of the costs associated with their own medical treatment via "out-of-pocket" expenses (OOPE). Other ways to pay for medical treatment include purchasing public or private health insurance, as well as contributing to the support of public medical institutions. This chapter provides an overview of the components of expenditures on healthcare as well as the need for and the resultant difficulties associated with, the assessment of the cost of providing healthcare services in the nation.

6.1 INTRODUCTION

In 1947, India declared independence from the British Empire, and the Ministry of Health was founded as a result. Every 5-year plan sets out the state's expenditure objectives for the next five years, and health is one of them. Parliament approved the National Health Policy (NHP) in 1983. Healthcare coverage was a goal of the policy, which was modified in 2002. India's healthcare system is mostly controlled by the states. Each state in India is obligated to provide healthcare for its citizens under the country's constitution. The National Rural Health Mission was started by the national government in 2005 to address the shortage of medical coverage in rural regions. The goal of this mission is to improve healthcare in India's poorest regions by concentrating resources on rural areas and impoverished states with deficient health facilities. The Indian healthcare system and its financial processes are described in detail by the authors.[1] In India, everyone has access to a public healthcare system. As a result, medical care in India has wide variations in both quality and availability. In terms of healthcare, the disparity between rural and urban regions and states may be enormous. Physician shortages are common in rural regions, and state differences imply that citizens of the poorest states, like Bihar, frequently have less access to good healthcare than residents of those that are comparatively more prosperous. While the federal government is responsible for administrative and technical services, state governments are in charge of healthcare and health education.[2] Since India's public healthcare system does not adequately cover most of its citizens, many Indians must resort to private healthcare providers, an alternative that is often out of reach for the country's poorest residents. India has a wide range of health insurance options available. However, the vast majority of Indians remain uninsured, and out-of-pocket charges account for a significant amount of the country's medical expenditures. Private hospitals in India, on the contrary, provide top-notch medical treatment for a fraction of the cost of similar facilities in more developed nations.[3] Medical tourists go to India because of this facet of the country's healthcare system. Ayurvedic medicine, for example, is a popular alternative treatment option in India. Students of alternative medicine are also drawn to India as a tourist destination. To get sophisticated medical care in India, international students can expect to use private hospitals rather than government-run facilities. When it comes to minor health issues, pharmacists may be an excellent source of information. Knowing the Indian healthcare system and following appropriate health and safety measures can assist in guaranteeing that your stay in India is both healthy and pleasant. Globalization's impact on developing countries has led to the introduction of new technologies in the private sector rather than the public sector, which has resulted in an overestimation of public health services. Although they were equipped to handle an epidemic, healthcare systems were ill-prepared to prevent it. The prevention of illnesses falls within the purview of the economy's long-standing public institutions. Even in the United States, where the vast majority of medical treatments are paid for privately, this is a common tendency. It has been a long-term trend in India that governmental policies and programs have concentrated primarily on providing curative treatment and personal preventative actions such as vaccination. The Indian public healthcare system has a three-tiered structure: (i) Traditional public health (TPH) focuses on preventing illness rather than treating it. Draining swamps and eliminating pests such as rats and mosquitoes are two examples of government involvement in activities that the market would not pay for since they are considered pure public goods. Additionally, there are services such as vaccination, access to potable water, and sanitization that have significant externalities. States must take steps to promote the consumption of "merit goods" such as healthy diets and discourage the use of "demerit goods" such as smoking. (ii) A big part of primary care is routine, low-cost care and most treatment is done outside of a hospital. (iii) Secondary care is known for its high costs, high levels of specialization, and status as the main inpatient care.

In India, the nation's government has prioritized spending its money on postsecondary education above basic education. Large sections of the money have already been spent to subsidize additional high-end tertiary medical services as well as medical education at lower levels. These tend to favor the middle and higher classes of society while at the same time taking away resources that might have been utilized to provide public health services. Image processing and computer vision advances in the healthcare industry must be investigated. Even though this field has grown tremendously in the past several decades, progress has not been made to the degree that one would expect. Thus, we have delved into healthcare utilizing image processing and computer vision, but sensor-based activity is also a highly essential topic.[4–15] According to the idea of collective action, every effort to deliver a public benefit would always run into problems in terms of carrying it out in an effective manner. The general populace of a nation has a strong preference for allocating public funds to private goods, such as medical treatment, rather than public goods, such as sanitary measures, which contribute to an overall improvement in the community's level of health. This issue is addressed by public health authorities in developed countries to ensure adequate funding for the public healthcare sector.[16] The Internet of Things (IoT) allows healthcare practitioners to be more attentive and proactive in their interactions with patients. Physicians may use data gathered from IoT devices to discover the optimum treatment approach for patients and achieve the intended results.[17–21] On the contrary, this issue causes developing nations to neglect the same essential basic services because they are not addressed by public health authorities. The nondemocratic regime of East Asia, which is part of the group of developing nations, has been successful in improving health outcomes by utilizing its limited resources on public health measures rather than spending them on providing advanced medical care. Because of this change in spending priorities, more money is being put toward improving public health as a whole.

6.2 RELATED WORK

Thanks to state-of-the-art technology, the Telecare Medicine Information System (TMIS) now provides patients with rapid and convenient healthcare services at their doorsteps. This design, on the contrary, introduces additional dangers and difficulties for the privacy, integrity, and safety of patients and the server as a whole. Using an authentication strategy is commonly regarded as the most effective technique for the TMIS to check the validity of patients and the server to prevent any resource misuse and malicious attacks. As a result, a number of different authentication techniques have been put forward. With their new authentication procedure, the authors have fixed every security problem as discussed in the study by Qiu et al. Authors also compare themselves to others. The new protocol is more secure and efficient than earlier protocols, as shown by comparisons with comparable protocols.[22] Recent years have seen an increase in the importance and popularity of disaster relief and healthcare supply chain studies. However, only a few articles have been written regarding the present challenges in healthcare and the catastrophe supply chain, particularly in cases of natural disasters. Research on healthcare and catastrophe supply chains is the subject of this study. The authors have analyzed the last 10 years' worth of research on the subject. Between 2005 and 2014, published articles on healthcare and disaster supply chain research fell into three broad categories: (1) natural disaster healthcare supply chain; (2) disaster supply chain; and (3) natural disaster healthcare supply chain operational management, information technology, inventory and control management, strategic management, and service management are some of the primary subjects covered in this section. In addition, the types of research methodologies include empirical investigation, case study, modeling and simulation, literature review, and conceptual and theoretical theory. Based on its findings, further study in these three areas will be directed.[23] A structured questionnaire was used to collect 404 samples from in-patients at 20 recognized Ayurveda hospitals in the northern section of Kerala, India's southern state, using the easy sampling approach. Women over the age of 40 made up the majority of those who took part in the survey. Multiple regression analysis and structural equation modeling are both methods that were utilized to come up with two different models of the relationships between the variables in the research, and these models were then compared using ANOVA and *t*-tests. Consumer perceptions of service quality and patient satisfaction differ according to socioeconomic characteristics, except for education, according to research. There followed an evaluation of how performance-based service quality affects patient satisfaction and the mediation model.[24] New pharmaceuticals and therapies will be discovered thanks to the IoT, a developing technology. Healthcare efficiency and quality offer high-potential attributes such as flexibility, adaptability, affinity, cost reduction, and speed. Using this technology, researchers can better understand the precise threats to security and privacy. This article focuses on an ontology-driven healthcare information system. In the proposed Ontology-based healthcare information system, the security and privacy issues are examined in detail. Ambulances, fire trucks, and other emergency vehicles are all examples of EMS units. A functional infrastructure design is also presented to demonstrate how the suggested application architecture is integrated with the IIoT and ontology hierarchy.[25] Using distant servers to transmit signals may speed up medical services in terms of both time and space. When it comes to large-scale medical projects and offering sensitive and intense services, latency reduction is an essential element of computing platforms that may allow the completion of healthcare procedures. One of the study topics relating to the provision of sensitive and intensive services is lowering the cost of transmitting data to the cloud. The research aims to reduce data delivery costs in the cloud.[26] Think about India's various challenges in building a sustainable and feasible strategy for healthcare tourism and all the other parties involved in providing cross-border access to healthcare to maintain its leading position in healthcare tourism. It is expected that India will be able to take advantage of the steady increase in medical tourism it has seen in recent years by making investments in infrastructure, including public and private healthcare systems, fostering accreditation, and encouraging medical technology and insurance sector innovation.[27] Many regulatory frameworks persist and function in a middle-income society with considerable market-based provisioning, as we have shown. In Maharashtra's private healthcare sector, we have identified regulatory centers that are not in the corridors of health ministries and statutory professional bodies but rather in the offices of regulatory capitalism: accreditation companies, insurers, platform operators, and consumer courts. Some regulatory functions are carried out by the state's judiciary, while others are carried out by a variety of industry actors, private organizations, and public insurers to advance their own interests in the sector. This role is typically organized around legislation, licenses, and inspections.[28] It is vital that both the central government and the state governments work together to develop the public healthcare system to guarantee that all citizens have access to high-quality treatment. The Pradhan Mantri Jan Arogya Yojana (PM-JAY) health insurance scheme, which is run by the central government, is an encouraging start toward meeting the requirements of disadvantaged demographic segments in terms of medical treatment.[29] According to the authors' research and information, having class disparity is the primary reason for having unfulfilled needs despite the minor influence of gender and caste. However, the interaction of economic class with caste and gender reveals even greater disparities in access to medical care. Although gaps in unmet needs across classes may be shown to exist across caste as well as gender, it is the caste inequalities that exacerbate the discrepancies that exist within classes. The removal of obstacles preventing access to healthcare necessitates the development of a multidimensional framework for the identification of combinations of characteristics that should be directed toward the achievement of universal health coverage. For healthcare to be fair, these findings need to be taken into account when coming up with new plans and reorganizing the ones that are already in place.[30]

6.3 CHALLENGES OF HEALTHCARE SYSTEMS

The government's actions in dealing with this current pandemic show that alignment between stakeholders, including local communities, scientific communities across a wide range of disciplines, and public and private medical and regulatory authorities, as well as even legal authorities, is essential for dealing with epidemics. Even before the outbreak of the pandemic, the Indian healthcare industry has undergone a number of transformations that have the potential to radically alter the relationships between the various players. The "NRHM" abbreviation stands for the "National Rural Health Mission," which began in 2005–2006 with the purpose of improving rural healthcare delivery by enlisting the help of Panchayati Raj Institutions (PRI) and a large number of Accredited Social Health Advocates (ASHA) professionals. Because of this change in responsibility, NRHM has made it possible for the community to become an important stakeholder in the design and implementation of healthcare services in their own communities. Indian public healthcare has seen a dramatic transition in recent decades, and this development has served as a model for other developing nations throughout the globe. Healthcare in India's urban and rural regions is vastly different, not because of rare illnesses but because of the absence of suitable facilities and health professionals, among many other problems in the country's healthcare system. A primary focus of this chapter; however, is to identify and discuss the present issues facing both the public and private healthcare sectors in Figure 6.1.

a. **Budgeting and Insufficient Infrastructure:** This discrepancy in administrations' money management should be acknowledged and balanced. There is a wide disparity between urban and rural healthcare services in India because of a lack of funding for public healthcare. Private healthcare is preferred by most rural residents owing to low income and the absence of basic insurance, making it unaffordable for the majority of the population. For a substantial segment of Indian society, access to high-quality private healthcare is out of reach due to its high cost. Basic infrastructure is lacking in the public healthcare system, despite the purchase of a high-priced medical technology.

b. **The Struggle to Provide an Optimal Service Based on Insurance:** The insurance sector in India is still severely under-penetrated because of a lack of understanding of insurance as a concept and the benefits it offers. There has been a tremendous increase in insurance coverage in India as a result of public awareness campaigns and Indian government programs such as Ayushman Bharat, Employment State Insurance Scheme, etc. Healthcare delivery and service issues for the public sector healthcare system arise when there are several centers and state-specific insurance plans.

c. **Reduced focus on Health Promotion and Disease Prevention:** Preventive care in India receives about as much attention as insurance does. In terms of pain or financial loss, preventive treatment may save the patient a lot of time and money. The present scenario is a direct result of a lack of knowledge. For patients and the country's limited healthcare infrastructure, proper preventative treatment may save a lot of money. Despite government initiatives to educate the public about the dangers of the COVID-19 virus, many individuals continue to violate lockdown regulations and fail to wear even the most basic protective gear, such as masks and gloves. General health advice may not be able to address the underlying cause of the disease and may result in higher treatment costs in the long run due to the need for more time and resources. The nation urgently needs more medical specialists with specialized training.

d. **Delayed Recognition of Disease:** When it comes to saving money and lives, a stitch in time saves nine the next step after preventative care is the diagnosis, which if done in time, may not only save money but also help save lives. On a regular basis, we hear stories of situations that were just recently discovered when it was too late to save the patient's life. In India, diagnosis is often seen as a waste of time.

e. **Medical Research has no Financial Backing:** A lack of investment in R&D and cutting-edge technology-driven new initiatives has kept Indian firms from taking advantage of global trends and becoming contract manufacturers for multinational corporations (MNCs). Even when it comes to keeping up with the latest medical breakthroughs, we lag far behind. To provide patients with the best care possible, healthcare providers and pharmaceutical firms alike must stay abreast of the most recent developments in the medical area. Modern technology comes at a high price. Medical gadgets manufactured in India that are more dependable and cost-effective may assist administrators in maintaining a balanced fund profile and save a significant amount of money that can be utilized to provide basic patient services.

f. **Transparency and Cost:** Most of us are concerned about how much therapy will cost and how that information will be disclosed. To avoid the patient being startled when they get their bill for services, it should be such as a menu card at a restaurant; ingredients and prices are clearly stated in advance. Aside from that, patients have the option of choosing whether or not to use these services based on their budget. In addition, patients may select whether or not to use these services, keeping their budget in mind.

Issues in the public and private healthcare sectors are shown in Figure 6.1.

There are significant issues facing this business as a result of the country's demographic changes, the bad quality of the public infrastructure, and a lack of financial and human resources. All of these issues stem from the public sector's appallingly low involvement in the healthcare business. As much as the NHP strives to address many of these issues, it falls short due to implementation difficulties and a lack of funding.

FIGURE 6.1 Issues in the public and private healthcare sectors.

6.4 CURRENT CONCERNS ABOUT MEDICAL TREATMENT IN INDIA

The current challenges with healthcare in India vary from concerns with infrastructure to glaring inefficiencies in the bureaucratic system used by hospital administrations. Let us have a look at some of the most significant ones, which are as follows:

a. **Infrastructural problems:** The absence of adequate infrastructure is by far the most significant challenge that the healthcare industry in India is confronted with at the present time. A national healthcare system that is overburdened to an impossible degree by an ongoing shortage of hospital beds, a lack of specialized staff to handle serious illnesses, and significant out-of-pocket financial expenditures. To make matters worse, the pace at which experts are being trained is very low in comparison to other nations. As a result, the burden placed on this faltering healthcare system is precariously balanced.

b. **Rates of death among children:** Despite the incredible scientific developments that have been made over the last two decades, the mortality rate in India is still quite high, while the rest of the globe is seeing a reduction in this statistic. There are seven lakh twenty-one thousand child deaths per year in India, which is one of the highest rates of infant mortality in the world, as stated in an article that was published by the *Times of India*. There is a widespread misconception that prenatal surgical operations and baby care are too expensive for families to afford, which is one of the key contributors to the persistently high death rates. What we need to do is educate people about low-cost initiatives that are equally as effective as more costly ones, and we need to do this as quickly as possible. In India, not only will this have the long-term effect of lowering death rates, but it will also help strengthen people's faith in the nation's healthcare system.

c. **An unmanageable number of patients:** People should be focused on ensuring that medical and human resources can be sustainably managed to meet the requirements of future generations at this time. Healthcare institutions should use technology to improve operational and clinical processes as much as possible to generate sufficient patient flow. Virtual care procedures and telehealth services, which both have the potential to reduce patient load stress, are hard to promote for reasons other than what is obvious.

6.5 SOLUTION FOR HEALTHCARE CHALLENGES

To address these issues, a new healthcare system must be developed instead of a mediocre replica of the one already in use across the globe. Building new hospitals, clinics, and medical schools as a typical policy strategy across many Indian states would be difficult, time-consuming, and costly, if not impossible, and although it may be popular in the short term, it will create issues in the long run. As previously noted by *Signs of the Times*, digital technology has the capacity to change industries such as financial services, government, and retail. Many special advantages of technology make it particularly well-suited for transforming healthcare as well, including mass scalability, direct access, flexibility, cost-effectiveness, and transparency.

It is worth looking at five current technologies that might potentially disrupt the current healthcare delivery system to see whether there is an opportunity for innovation.

i. **Telemedicine:** Telemedicine is the use of telecommunications technology by medical experts to assess, diagnose, and treat patients at a distance Figure 6.2. In the past decade, the method has undergone a remarkable transformation and is now playing an increasingly crucial role in American healthcare.

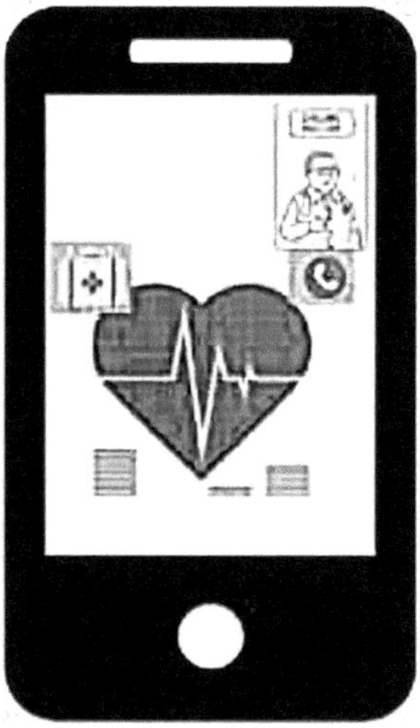

FIGURE 6.2 Telemedicine.

Telemedicine is a broad term that refers to any method in which you and your doctor can communicate virtually rather than face-to-face. It encompasses all forms of communication, such as voice conversations, video chats, email, and SMS. Telehealth, digital medicine, e-health, and m-health are other terms used to describe it. If your doctor provides telemedicine, all you need is a phone, smartphone, or computer with a stable internet connection. A useful tool for everyone, telemedicine is particularly beneficial for those who live in remote areas, have restricted mobility and transportation, or need medical treatment when they are away from home. Medical care may be obtained in a variety of ways depending on the services provided by your physician. These include the following:

a. *A website for patients:* Send and receive emails from your doctor or nurse, seek refills for your medication, and arrange appointments through a patient portal. Additionally, your doctor may explain to you the relevance of your lab or imaging test results. If you do not have the time to wait for a phone call, this is a better option.

b. *Making appointments with a virtual assistant:* It is possible to make an appointment with a doctor by phone or videoconference. These sessions are often held in places such as urgent care clinics and clinics for mental and behavioral health.

ii. **Apps and Digital Health Information:** There is also the possibility that mobile health applications may make healthcare more cost-effective. They make it possible to do remote monitoring, therapy, and surgery, as well as remote medication management, which may assist in cutting down on travel time and costs for both patients and clinicians. However, to accomplish this goal, it is necessary for individuals to have faith that the health apps they use are capable of gathering and analyzing data in a secure and efficient manner, in accordance with stringent data standards and regulatory oversight, and most importantly, that clinicians are taking the results into consideration.

iii. **Big Data and Analytics:** The medical and data communities are working together throughout the world to enhance the precision with which physicians are able to anticipate epidemics, treat illnesses, and improve the overall quality of life of people. This is being accomplished via the use of big data analytics in Figure 6.3. It seems that the first steps of this process are already in place in India: Wockhardt is one of India's largest hospital chains. It recently teamed up with QlikTech, a data and analytics company, to design and implement a business intelligence system that will help its doctors better analyze patient data and offer more affordable and personalized healthcare solutions.

FIGURE 6.3 Big data analytics in healthcare.

iv. **Wearable Technology:** Wearable technology in the healthcare industry refers to electronic gadgets that customers may wear, such as smartwatches and Fitbits. These types of devices are intended to gather data on the users' health and the amount of activity they get in Figure 6.4. These gadgets even have the ability to transmit real-time transmissions of a user's health information to a doctor or other healthcare expert. The number of customers who are interested in the idea of sharing the data collected by their wearable devices with their healthcare providers and insurers is expected to increase over the next few years, which will result in an increase in the demand for wearables.

FIGURE 6.4 Wearable technology in healthcare.

v. **Training in Medicine Using Virtual Reality:** Despite the fact that virtual and augmented reality technology has existed for some time in the form of simulators, the prohibitively costly gear and processing requirements make it inaccessible for widespread usage. Virtual and augmented reality are rapidly being used by governments and organizations to teach staff in a broad variety of professions because of the recent development of low-cost technology that is compatible with smartphones. Virtual reality is rapidly being utilized in healthcare to teach physicians in real-time scenarios and surroundings. So far, Medical Frontiers, a UK-based healthcare technology business, has trained 14,000 cancer surgeons in 113 countries using Google Glass VR.

6.6 CONCLUSION AND FUTURE SCOPE

Health systems and policies play an important role in determining how health services are delivered, and utilized, and influence health outcomes. Healthcare in India is failing, yet there is a great deal of promise and development in the country's healthcare education. Aside from addressing the apparent infrastructure issues, professional training should be prioritized. The use of online healthcare workshops and courses offered by reputable platforms enables more sophisticated and specialized training. Good quality medical training is priceless and investing in this area will benefit us in the long term. India's healthcare system consists of government-run institutions, private clinics, corporate hospitals, and unlicensed providers of healthcare. It is a combination of direct out-of-pocket payments by patients, government subsidies, donations, charity organizations, and pre-paid full or partial health insurance that is used to pay for healthcare. Taking this into account while doing a cost analysis, it is vital to account for the kind of facility, the services given, and how services are funded.

However, there are a few features that we can anticipate in India's healthcare system in the future.

a. Since the cost of healthcare in India is lower than in Western Europe or the United States, international patients flock to the country to get high-quality treatment. There will be a significant increase in medical tourism in the near future because of the outstanding vaccine supply agreements with stable economies such as Russia and Brazil as well.
b. Online consultations and technical platforms are in great demand across today's globe. Since the epidemic has increased the requirement for social distancing, teleconsultations have become a necessity rather than a desire for most people. By bringing together India's cities and villages, there may be a better way to share healthcare.
c. In a positive development for India, healthcare insurance, one of the country's most pressing health issues, has made substantial progress. There has been an increase in awareness of health insurance products in recent years, and the number of individuals acquiring them is rising year after year.

KEYWORDS

- **healthcare**
- **public healthcare**
- **India**
- **private healthcare**
- **future-ready**
- **IoT**
- **medical research**

REFERENCES

1. Shamsuddin, S.; Akkawi, M. E.; Zaidi, S. T.; Ming, L. C.; Manan, M. M. Antimicrobial Drug use in Primary Healthcare Clinics: A Retrospective Evaluation. *Int. J. Infect. Dis.* **2016,** *52,* 16–22. https://doi.org/10.1016/j.ijid.2016.09.013
2. Zeshan, F.; Mohamad, R. Medical Ontology in the Dynamic Healthcare Environment. *Procedia Comput. Sci.* **2012,** *10,* 340–348. https://doi.org/10.1016/j.procs.2012.06.045
3. Wang, Y.; Yang, J.; Qiao, F.; Feng, B.; Hu, F.; Xi, Z.; Wu, W.; Ni, Z.; Liu, L.; Yuan, Y. Compared Hand Hygiene Compliance among Healthcare Providers before and after the COVID-19 Pandemic: A Rapid Review and Meta-Analysis. *Am. J. Infect. Control* **2022,** *50* (5), 563–571. https://doi.org/10.1016/j.ajic.2021.11.030
4. Ahad, M. A.; Kobashi, S.; Tavares, J. M. Advancements of Image Processing and Vision in Healthcare. *J. Healthc. Eng.* **2018,** *2018,* 1–3. https://doi.org/10.1155/2018/8458024
5. Kumar, N.; Shukla, H. S.; Tiwari, A. K.; Dahiya, A. K. Dual Ascent Based Median Filter for Image Restoration. *SSRN Electron. J.* **2019**. https://doi.org/10.2139/ssrn.3350334.
6. Kumar, N.; Kumar, A.; Dahiya, A. K. Image Restoration Using a Fuzzy-Based Median Filter and Modified Firefly Optimization Algorithm. *Int. J. Adv. Sci. Tech.* **2020,** *29* (4s), 1471–1477.
7. Shukla, H. S.; Tripathi, R. P.; Kumar, N. Gaussian Noise Filtering Techniques using New Median Filter. *Int. J. Comput. App.* **2014,** *95* (12), 12–15.
8. Kumar, N.; Tiwari, A.; Dahiya, A. K.; Shukla, H. S. In *A Dual Ascent-based Median Filter for Image Restoration*, International Conference on Advanced Computing and Software Engineering (ICACSE-2019) 2019, vol 2, pp 432–435.
9. Pandey, H.; Mishra, A.; Kumar, D. Various Aspects of Sentiment Analysis: A Review. *SSRN Electron. J.* **2019**. https://doi.org/10.2139/ssrn.3349011

10. Kumar, N.; Dahiya, A. K.; Kumar, K. Modified Median Filter for Image Denoising. *Int. J. Adv. Sci. Tech.* **2020,** *29* (4s), 1495–1502.
11. Shukla, H. S.; Kumar, N.; Tripathi, R. P. Median Filter-based Wavelet Transforms for Multilevel Noise. *Int. J. Comput. App.* **2014,** *107* (14).
12. Kumar, N.; Shukla, H. S.; Tripathi, R. P. Image Restoration in Noisy Free Images using Fuzzy-based Median Filtering and Adaptive Particle Swarm Optimization - Richardson-Lucy Algorithm. *Int. J. Intell. Eng. Syst.* **2017,** *10* (4), 50–59.
13. Shukla, H. S.; Tripathi, R. P.; Kumar, N. Image Restoration using Modified Binary Particle Swarm Optimization Richardson-Lucy (MBSO-RL) Algorithm. *Int. J. Appl. Eng. Res.* **2015,** *10* (22), 43077–43081
14. Kumar, K.; Kumar, N.; Shah, R. Role of IOT to Avoid Spreading of Covid-19. *Int. J. Intell. Networks* **2020,** *1,* 32–35. https://doi.org/10.1016/j.ijin.2020.05.002
15. Kumar, N.; Kumar, S.; Dahiya, A. K.; Kumar K. In *Application of IoT in Agriculture*, 9th International Conference on Reliability, Infocom Technologies and Optimization (Trends and Future Directions) (ICRITO) 2021, vol 2021, pp 1–4.
16. Faqurudheen, H.; Mathew, S.; Kumar, T. M. Exploring Family Resilience in a Community Mental Health Setup in South India. *Procedia Econ. Financ.* **2014,** *18,* 391–399. https://doi.org/10.1016/s2212-5671(14)00955-1
17. Kumar, K.; Kumar, A.; Kumar, N.; Mohammed, M. A.; Al-Waisy, A. S.; Jaber, M. M.; Shah, R.; Al-Andoli, M. N. Dimensions of Internet of Things: Technological Taxonomy Architecture Applications and Open Challenges—A Systematic Review. *Wirel. Commun. Mob. Comput.* **2022,** *2022,* 1–23. https://doi.org/10.1155/2022/9148373
18. Kumar, N.; Kumar, K.; Kumar, A. In *Application of Internet of Things in Image Processing*, 2022 IEEE Delhi Section Conference (DELCON), 2022. https://doi.org/10.1109/delcon54057.2022.9753308
19. Kumar, N.; Dahiya, A. K.; Kumar, K.; Tanwar, S. In *Application of IoT in Agriculture*, 2021 9th International Conference on Reliability, Infocom Technologies and Optimization (Trends and Future Directions) (ICRITO), 2021. https://doi.org/10.1109/icrito51393.2021.9596120
20. Kumar, K.; Saini, G.; Kumar, N.; Kaiser, M. S.; Kannan, R.; Shah, R. Prediction of Energy Generation Target of Hydropower Plants using Artificial Neural Networks. In *Sustainable Developments by Artificial Intelligence and Machine Learning for Renewable Energies*; 2022, pp 309–320. https://doi.org/10.1016/b978-0-323-91228-0.00005-7
21. Kumar, K.; Singh, R. P.; Ranjan, P.; Kumar, N. Daily Plant Load Analysis of a Hydropower Plant using Machine Learning. *Algorithms Intelligent Syst.* **2021,** 819–826. https://doi.org/10.1007/978-981-33-4604-8_65
22. Shamshad, S.; Ayub, M. F.; Mahmood, K.; Kumari, S.; Chaudhry, S. A.; Chen, C.-M. An Enhanced Scheme for Mutual Authentication for Healthcare Services. *Digit. Commun. Netw.* **2022,** *8* (2), 150–161. https://doi.org/10.1016/j.dcan.2021.07.002
23. Syahrir, I.; Suparno, V. I. Healthcare and Disaster Supply Chain: Literature Review and Future Research. *Procedia Manuf.* **2015,** *4,* 2–9. https://doi.org/10.1016/j.promfg.2015.11.007
24. Suhail, P.; Srinivasulu, Y. Perception of Service Quality, Satisfaction, and Behavioral Intentions in Ayurveda Healthcare. *J. Ayurveda Integr. Med.* **2021,** *12* (1), 93–101. https://doi.org/10.1016/j.jaim.2020.10.011

25. Rinty, M. R.; Prodhan, U. K.; Rahman, M. M. A Prospective Interoperable Distributed E-health System with Loose Coupling in Improving Healthcare Services for Developing Countries. *Array* **2022,** *13,* 100114. https://doi.org/10.1016/j.array.2021.100114
26. Mani, N.; Singh, A.; Nimmagadda, S. L. An IoT Guided Healthcare Monitoring System for Managing Real-Time Notifications by Fog Computing Services. *Procedia Comput. Sci.* **2020,** *167,* 850–859. https://doi.org/10.1016/j.procs.2020.03.424
27. Jadhav, S.; Yeravdekar, R.; Kulkarni, M. Cross-Border Healthcare Access in South Asian Countries: Learnings for Sustainable Healthcare Tourism in India. *Procedia Soc Behav. Sci.* **2014,** *157,* 109–117. https://doi.org/10.1016/j.sbspro.2014.11.014
28. Hunter, B. M.; Murray, S. F.; Marathe, S.; Chakravarthi, I. Decentred Regulation: The Case of Private Healthcare in India. *World Dev.* **2022,** *155,* 105889. https://doi.org/10.1016/j.worlddev.2022.105889
29. Yadav, J.; Devi, S.; Singh, M. N.; Manchanda, N.; Moradhawaj. Healthcare Utilization and Expenditure Inequities in India: Benefit Incidence Analysis. *Clin. Epidemiol. Glob. Health* **2022,** *15,* 101053. https://doi.org/10.1016/j.cegh.2022.101053
30. Mahapatro, S. R.; James, K. S.; Mishra, U. S. Intersection of Class, Caste, Gender and Unmet Healthcare needs in India: Implications for Health Policy. *Health Policy Open* **2021,** *2,* 100040. https://doi.org/10.1016/j.hpopen.2021.100040

CHAPTER 7

IoT-BASED ADVANCED HEALTH-MONITORING SYSTEM USING CLOUD PLATFORM

NARENDRA KUMAR[1], KRISHNA KUMAR[2], and INDRADEO RAM[3]

[1]*School of Computing, DIT University Dehradun, Dehradun,Uttarakhand, India*

[2]*Department of Hydro and Renewable Energy, Indian Institute of Technology, Roorkee, India*

[3]*Department of Computer Science and Engineering, BIT Mesra, Ranchi, Jharkhand, India*

ABSTRACT

The Internet of Things (IoT) is a new and rapidly evolving technology in which everything (smart items and smart gadgets) is linked to the internet to allow for efficient communication between these interconnected devices. The Internet of Things is a stimulant for healthcare and plays a critical role in a broad variety of healthcare-monitoring applications. Networked sensor devices, whether worn on the body or implanted in living situations, enable the collection of rich information to assess the patient's overall health state by collecting body temperature, blood pressure, sugar level, and so on. The difficult challenge with the Internet of Things is communicating this acquired information to the doctor, making appropriate decisions based on the data obtained, and alerting the patient. In this paper ESP8266 and GPS track the patient's current location. We use ESP8266 WiFi Board as

IoT-Enabled Healthcare Systems: Applications, Benefits, and Challenges
Yatindra Kumar, Shashi Kant Verma, Archana Singh, Krishna Kumar, & Manoj Gupta (Eds.)

a processor where the collected data is sent to ESP8266 and transmit data over the Internet for analysis. This analyzed data is saved and may be utilized in a variety of ways. When a serious situation is recognized and the alarm system is triggered, the health data is immediately transferred to the doctor. ThingSpeak generates real-time visualizations of data sent by your devices. We can perform online monitoring and processing of data as it comes in using ThingSpeak.

7.1 INTRODUCTION

The epidemic of COVID-19 has compelled researchers to use telemedicine to deliver trustworthy and secure medical services remotely. The doctor has to know the current state of the patient's health in order to have a better picture of the patient's predicament. The twentieth century saw a significant shift in technology. The Internet has grown such that resources and intelligent objects may be accessed from anywhere, at any time. For decades, automation and wireless technologies have played a significant part in the Internet of Things (IoT).[14–16] Network capacity will be fully used by the Internet of Things (IoT), changing the deployment of disruptive services in large-scale scenarios such as smart cities and healthcare. Industry 4.0 has arrived, which implies that humans are now attempting to communicate with machines. Using a microcontroller, an influence range, communication interface, and a collection of actuators and sensors to interact with the environment, IoT applications may benefit from a wide distribution and analysis of sensitive item installations.

a. ***CGM (continuous glucose monitoring):***

   A continuous glucose monitor was authorized by the FDA in 1999 (CGM). The CGM is a gadget that takes frequent readings of blood glucose levels in order to provide continuous monitoring. To use a CGM, a small sensor is inserted under your skin, often on your abdomen or arm. When it comes to measuring glucose levels in the blood, the sensor measures interstitial glucose. The sensor checks the glucose level in the sample in every few minutes. The data is sent to a monitor through a wireless transmitter. An insulin pump may have a display, but it may also be a separate item that you carry in your handbag or pocket. The data from certain CGMs may be sent immediately to a tablet or smartphone through

Bluetooth or USB. Glucose levels are monitored with a CGM, a small sensor that is inserted under the skin. As seen in Figure 7.1, information is sent from a transmitter to a receiver.

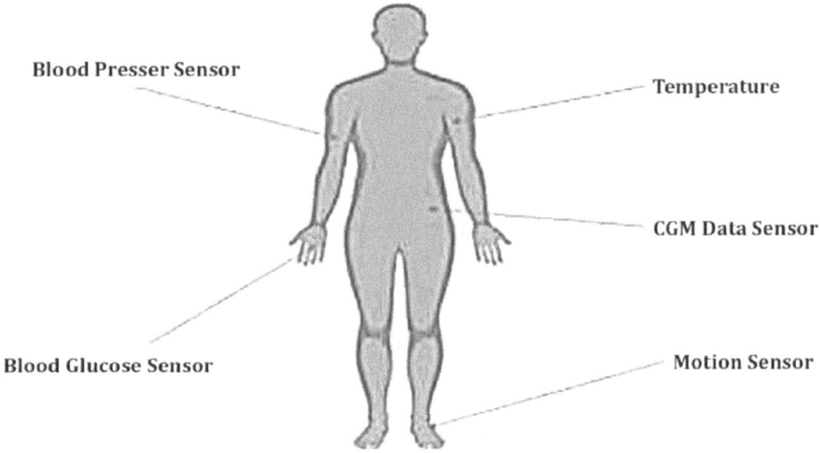

FIGURE 7.1 Sensors in human body.

b. ***Inhalers that are linked:***

There are hundreds of millions of individuals throughout the globe who suffer from asthma. Their symptoms and treatments may be better managed with the use of smart technology. With its smart inhaler technology, Propeller Health is the world's most prominent producer. As an alternative to generating entire inhalers, Propeller developed an attachment for the Bluetooth spirometer or an inhaler that connects to a smartphone application and assists patients with asthma, including emphysema, chronic bronchitis, and COPD.

c. ***Contact lenses that are linked:***

In 2014, Google Life Sciences stated that it would be developing an early warning system for diabetics to inform them when their blood glucose levels decrease or increase over a certain threshold, which would allow them to take preventative measures. Alcon, Novartis's eye care subsidiary, was enlisted to help with the endeavor.

d. ***Asthma Monitor ADAMM:***

ADAMM is a wearable, cutting-edge asthma gadget that anticipates the onset of an attack and alerts the user so that they may take action before things get out of hand. When used in combination with an app and a web-based interface, ADAMM makes it easier for asthma sufferers to send out medication alerts.

e. ***Pacemaker Monitoring that is Wireless and Automated:***

In today's pacemakers, wireless connectivity to external control systems is common. In this way, patients will no longer be required to attend the hospital only to have their medical gadgets evaluated. With these systems, patients and medical devices may be monitored on a day-to-day basis. Using remote interrogations, doctors can keep tabs on patients between their normal hospital visits and may warn them of any potential problems.

7.2 LITERATURE REVIEW

Artificial heart pacemakers were the subject of research by Kulkarni et al. IoT security challenges were handled by Tawalbeh et al.[2] by utilizing a variety of security models, including attribute-based access control (ABAC) on the Amazon Web Service (AWS) cloud provider platform. As presented by Yuan et al.[3] two-stage computation offloading potential game (TPOS) approach was developed to maximize resource allocation while also taking into account the WBAN tasks and the users' preferences. A system utility maximization issue was first constructed using the quality of service (QoS) of jobs. The offloading of computations may be modeled using the reward, cost, and penalty functions. Reciprocal constraint strategies in the strategy space of the future game model necessitated the development of a two-stage optimization technique to reduce computing time and increase algorithm feasibility. During efficiency testing, it was shown that the TPOS algorithm had a significant impact on average processing time, energy consumption, and network utility during efficiency testing.

According to Alam et al.,[4] finger and temperature sensors were used to create a microcontroller-based heartbeat and body temperature monitor. In contrast to standard recording devices, the instrument employs optical technology to monitor blood flow via the finger. Hussen and colleagues[5] studied rural Healthcare monitoring systems because of the low doctor-to-patient ratio in rural locations. Patients' essential physiological data may be tracked via wearable devices and mobile apps, according to ElNour et al.[6] Patients and Healthcare professionals may also be tracked using GPS technology. Smart health-monitoring systems based on the Internet of Things were studied by Rahaman et al.[7] For the IoT-based smart health-monitoring system, the pros and demerits of the new revolutionary technologies were considered.

A WiGym, a passive Wi-Fi monitoring gadget, was proposed by Liu et al.[8] to track and record a user's fitness activities. There is a distinct pattern in the time series of CSI reflected multipaths toward the body due to the body's movements. To keep track of how many workouts a person completes, short-term energy is taken into account. Functions are removed from the CSI time series using a discrete wavelet transform. To identify free-weight and body-weight activities, Mortazavi et al.[9] developed an extraction and selection function that uses the workout routine backdrop to generate effective classifiers. Meivel et al.[10] proposed and executed an E-health programme. Sensors included throughout the earth's architecture capture human actions and movement. Indicators of changing well-being, such as design alterations, have been found.

The goal of Qi et al.[11] was to identify exercises in the gym (GPAs). For free weight workouts, there is a two-layer identification system that can count repetitions as well as sets for free weight exercises. As a preliminary step, an arbitrary distinction between free weight and non-free weight activities of aerobic and sedentary activity identification is made using a single-class SVM (OC-SVM). This layer uses a neural network (NN) and a HMM (hidden Markov model) for further categorization into activities with no weight. Extraction of the PPG signal from the measured PPG using the sparse approach and then HR estimation are two separate stages suggested in Thomas et al.[12] In the sparse technique, signals are correctly reconstructed from a small number of noisy signals. This method does away with the requirement to read the accelerometer and instead utilizes two PPG signals, making it easier.

7.3 SYSTEM ARCHTECTURE

Layers one through four of the protocol stack are the physical, middle-ware, network, and application. To begin with, the physical layer contains the devices that have sensors and transmitters incorporated. It is up to the middleware layer to store data in the cloud, and it is also up to the network layer to transmit signals from inbuilt sensors to cloud. Finally, the application layer is responsible for analytics and diagnostics. A circuit diagram of the proposed architecture is shown in Figure 7.2.

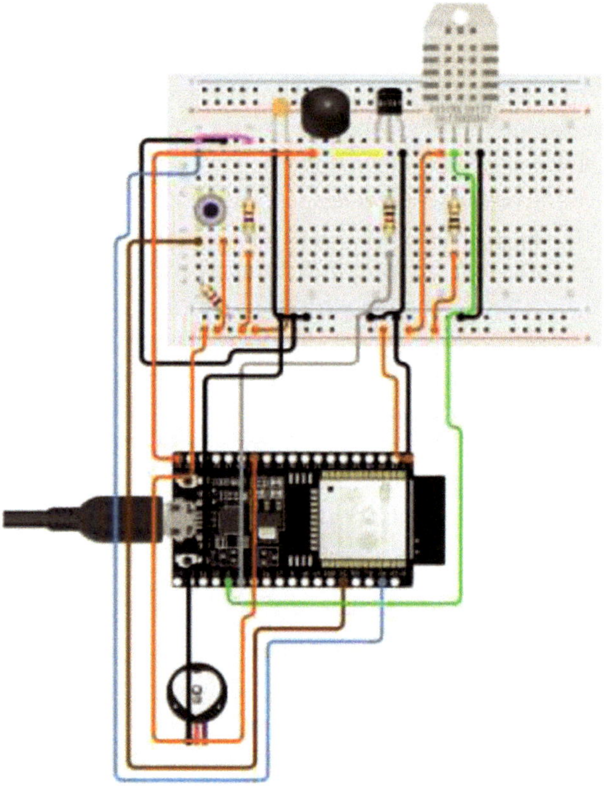

FIGURE 7.2 Circuit diagram.

Patients will be given wearable sensors that can measure electrocardiography (ECG), temperature, electromyography (EMG), muscle activity, sweating, respiration rate, and blood glucose level. Diseases such as arrhythmia, neuromuscular abnormalities, fever, obesity, blood pressure, and diabetes can be treated using these devices. Sensors that can be conveniently placed in contact with the skin in several body areas are highly preferred for obtaining reliable measurements.

A variety of physiological parameters, including heart rate, temperature, muscular activity, respiration rate, perspiration, and blood glucose level, will be monitored via sensors worn by patients. Arrhythmia, fever, neuromuscular disorders, blood pressure, obesity, and diabetes may all be treated with these gadgets, as can a host of other conditions as shown in Figure 7.3. Sensors that may be put directly on the skin in various parts of the body are recommended for accurate readings.

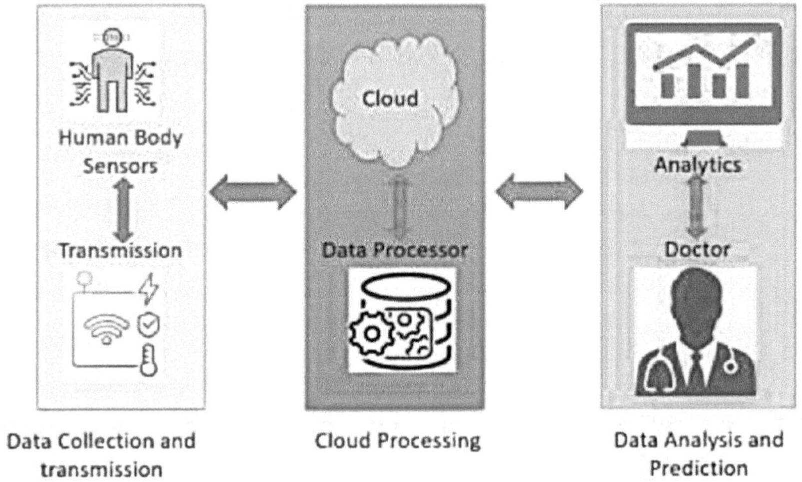

FIGURE 7.3 Architecture of IoT for health monitoring.

There are little sensors put in the patient's body that collect a wide range of physiological data. Next, a small piece of hardware may preprocess the data and a communication application can send the data out. A patient's movement or motions must be unaffected by the sensors' small size and low weight. Batteries that are small and low-energy are needed to power these sensors. It is expected that the batteries will not need to be recharged or replaced for an infinite period of time.

Patient records must be correctly and securely sent to the health center by system components in responsibility of data transmission.

It is possible to transmit short-range digital radio Zigbee or Bluetooth. Additionally, the information gathered may be sent to a medical facility for storage through the Internet. The concentrator, which may even be a smartphone, can operate the sensors in an IoT system over the Internet

a. **Cloudlet Processing:**

Nowadays, smartphones have more sophisticated capabilities that enable them to be used as both LTE and WiFi at the same time. Such smartphones may be used as concentrators in this configuration. Data from the concentrator will be stored in the cloud. For physicians, accessing this data on demand or using it for analytics would be immensely beneficial if this data is kept. In situations when local resources are inadequate, a cloudlet may be used to both store and process data locally. This tool also helps to perform time-sensitive tasks on the medical records of patients. Cloudlet storage makes data accessible at all times for data analytics, which may then be used to provide more precise diagnostics.

Cloudlet-computing has been proposed as a better option for PAN-based health applications since they commonly deal with offline data. Cloudlet and concenter may link through WiFi to reduce data transfer latency for critical operations involving data acquisition. Lastly, the data in the cloudlet will be retained in the cloud for trustworthy storage and dispersed data access, as previously stated. This kind of aggregation may be characterized by context aware focus, which refers to the patient's current and expected state of affairs.

b. **Analysis and Prediction:**

Data analysis is a key challenge in medical databases due to its size. A machine learning method establishes the relationship between sensor features and clinical data. Longer-term analysis of medical diagnostic data may boost its accuracy. Pattern recognition and machine learning methods will be used to wearable sensor data. Sensor data is becoming more diverse and dynamic, and machine learning must be updated to handle these challenges. Sensor designs vary often, therefore algorithms must be able to handle missing data values as well as data streams that are constantly changing in terms of their dimensions and semantics.

We face three significant issues when integrating IoT in the medical industry when it comes to analytics. There are a lot of new measuring tools and equipment in the area of medicine that are being launched on a regular basis. As a consequence, IoT devices and sensor data will need to be updated on a frequent basis.

IoT devices must be able to handle all of this, and the database architecture will have a significant influence. Because of the rapidly shifting nature of sensory data, it is projected that machine learning algorithms will continue to improve in the coming years. Second, the data to be gathered will vary based on the patient's condition, as ordered by the physician. As a result, it is difficult, if not impossible, to add new input in the future. Because of unusual patient situations, it is feasible to correlate past sensor data with clinical records, although the process is still difficult. Machine learning algorithms may benefit from classification and regression techniques, which can be used to create the common training data. However, this will place an extra strain on the doctors.

7.4 PROPOSED IoT-BASED ADVANCED HEALTH-MONITORING SYSTEM

When linked to the Internet of Things gateway, smart IoT sensors like Wi-Fi pacemakers and other devices may be processed in the cloud via cloud-based services. Smartwatches may monitor pulse rate, ECG monitoring, calorie intake/exhaustion, and body temperature. In order to transfer the data to the cloud, a CGM and an Asthma inhaler sensor will be used. Data from the sensor is sent via the gateway, which uploads it to a central server where it may be evaluated and monitored by medical professionals. The application of machine learning and online monitoring will allow for a more in-depth study of the data. Using data examined by the doctor, he or she may quickly determine the current health state of a patient and prescribe medication appropriately. IoT-based enhanced health-monitoring systems are shown in Figure 7.4.

The system's flow diagram is shown in Figure 7.5. Monitoring occurs as soon as the device's power is switched on and it sends data to the Arduino Uno, the Node MCU, and the controller. The Node MCU transmits the measured value to the stationary server. It warns both doctor and patient if oxygen saturation is less than 95% or pulse rate is 60 or 90 beats per minute or above. The system then displays this data, along with SpO2. The measured value may be seen through a mobile app or the device's LCD display, depending on the preferences of the user.

Data gathered by the microcontroller is not shared with other devices. In order for the microcontroller to carry data from one place to another

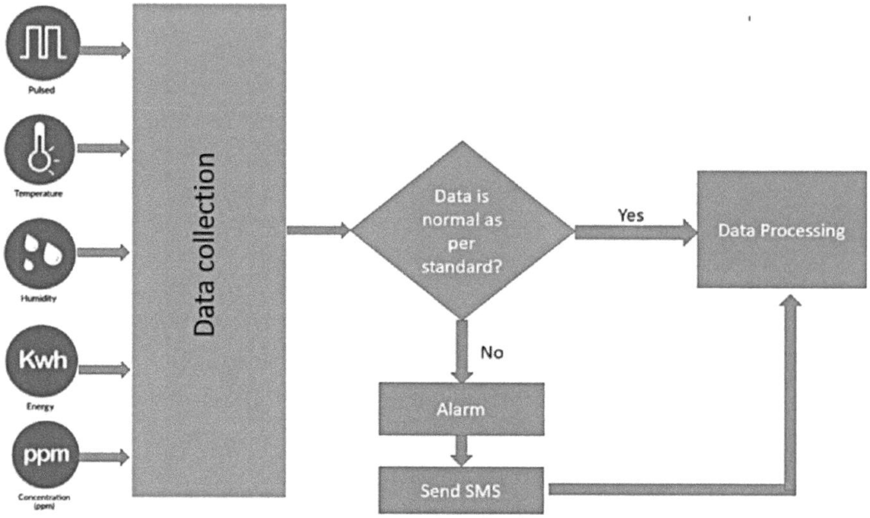

FIGURE 7.4 IoT-based enhanced health-monitoring systems.

at varying intervals, ESP8266 offers Wi-Fi services. The microcontroller code that we develop is used to connect the microcontroller to our devices. We may utilize n inputs, establish their threshold values, and mark them up in things speak cloud so that data obtained from the microcontroller can be analyzed properly. Additionally, we can see our data represented graphically, making it accessible to non-specialists like physicians as well as the general public. In order to produce a health analyzer that is accessible and understandable to everyone, this module has been designed.

As a last point, when we receive information from a number of sources, the sensory data will present several modalities. This heterogeneity is a barrier to machine learning algorithms that work with homogenous data. Many different input data may be combined in an efficient and flexible way using graphic models. Despite the fact that the sensor data is numerical, the medical data is shown graphically so that the patient's health may be constantly monitored. Health monitoring relies heavily on the concept of visualization. Several graphical methodologies are used to disseminate data from IoT warble sensors for good prediction. Emergency scenarios need speedy and accurate predictions using visualization tools that can handle a wide range of data. To compare patient medical records, the visualization must be able to handle static images.

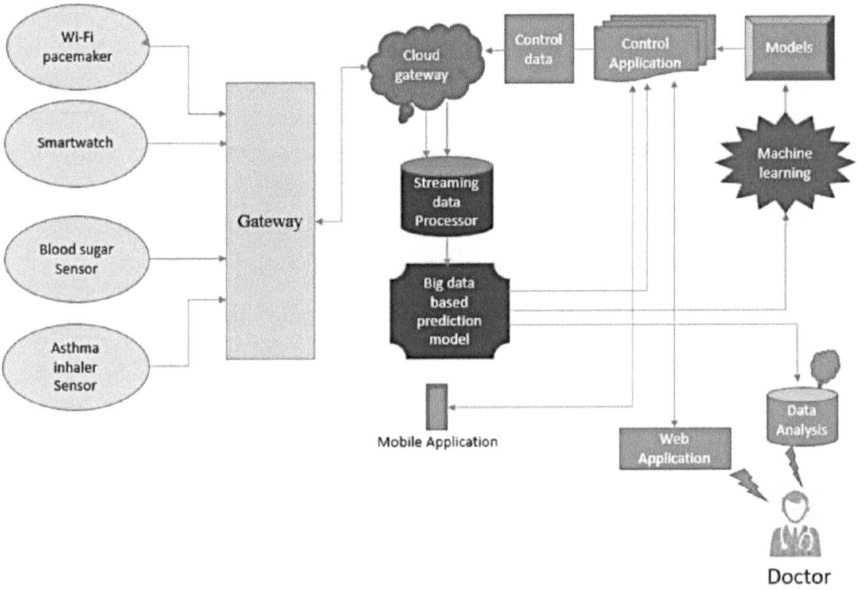

FIGURE 7.5 Proposed architecture for IoT-AHMS.

7.5 RESULT AND DISSCUSSION

People of many ages and backgrounds were used to test the system's performance. Heart rate, core body temperature, and ambient temperature were all monitored in these test cases, with the results manually calculated. In this example, the room temperature sensor is only used to measure humidity.

Person	Age	SpO2 (%)	Pulse (bpm)	Temperature (°C)
A	27	95	76	35
B	34	96	75	39
C	37	95	73	38
D	58	95	71	38
E	26	94	77	39

An Internet connection is all that is needed for a member of the medical team to keep tabs on a single patient and the current status of their room. A password-protected system ensures that only authorized personnel have access to the patient data at this facility. A doctor may easily make a decision from a distant location by looking at data. In a smart health environment, these indicators have a particular threshold. There is a point at which the medical team can easily implement the necessary healthcare operations.

Patient data such as pulse rate, temperature, and ECG values are recorded and are flexible enough to be seen on both LCDs and can be analyzed in Think speak website, along with credentials data, and a GSM module is attached to the project so that a message is sent to the respective number if the pulse rate falls below 70 and the temperature rises above 38. The real-time data of parameters are shown in Figure 7.6.

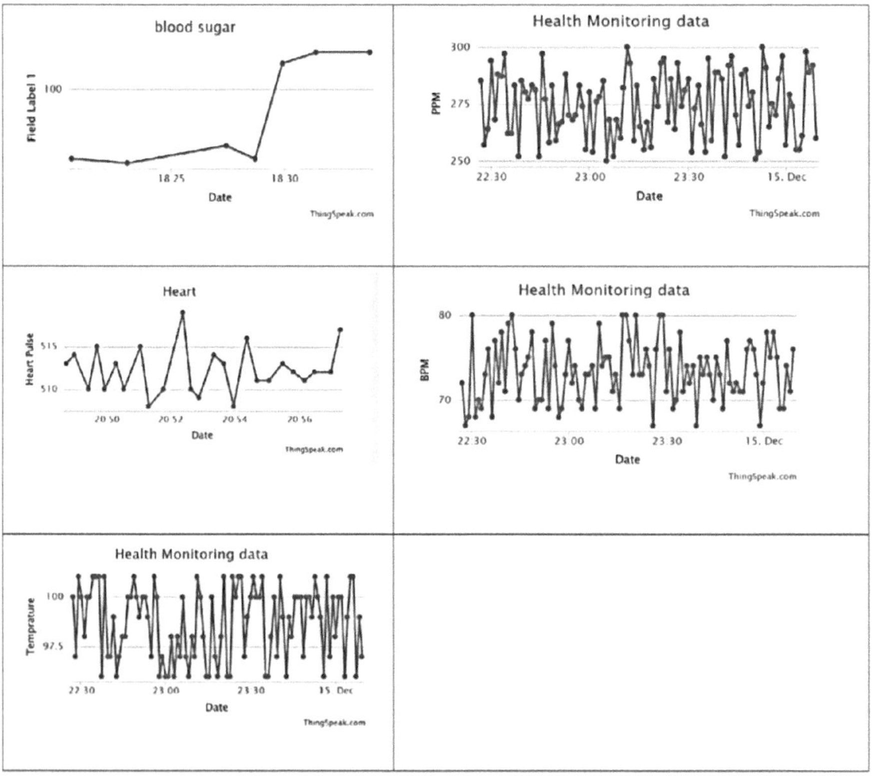

FIGURE 7.6 Realtime patient health data received from IoT sensors.

Health concerns may be monitored with the Remote Health Monitoring project module. It helps everyone in their daily health awareness practise. Anyone, from a young infant to an old person, may have their health assessed using this method. This module reduces the amount of time patients spend in the hospital as well as the number of times they need to attend.

7.6 CONCLUSION

The Internet of Things (IoT) is becoming more popular in the healthcare industry, particularly for the monitoring of patients with chronic conditions. Thanks to the new monitoring system's design, a doctor will be able to see in real time how a patient is doing. This allows doctors to quickly assess the patient's current condition and give medication based on real-time data. Because of the current COVID-19 outbreak, all hospital outpatient departments are now closed. The development of a telemedicine application is required to limit the danger of coming into contact with infected individuals. As a result of the IoT and tiny sensors, patients will be able to minimize their fear of danger even if they are away from home and their physician. Sensory data may be gathered in a variety of settings, including at home or at work. These obstacles in sensing, analytic, and illness prediction may be overcome in order to enable a smooth integration into the medical industry.

KEYWORDS

- **health-monitoring system**
- **ThingSpeak**
- **Internet of things**
- **cloud processing**
- **data analysis**

REFERENCES

1. Kulkarni, A.; Vijaykumar, S. Application of Internet of Things in Artificial Heart Pacemakers and Its Impact on Security. *Int. J. Curr. Trends Eng. Res.* **2016,** *2,* 604–610.
2. Tawalbeh, M. Authorization Model for IoT Healthcare Systems: Case Study 2020, 337–342.
3. Yuan, X.; Tian, H.; Wang, H.; Su, H.; Liu, J. Edge-Enabled WBANs for Efficient QoS Provisioning Healthcare Monitoring: A Two-Stage Potential Game-Based Computation Offloading Strategy 2020, 4. https://doi.org/10.1109/ACCESS.2020.2992639.

4. Alam, M. W.; Sultana, T.; Alam, M. S. A Heartbeat and Temperature Measuring System for Remote Health Monitoring Using Wireless Body Area Network. *Int. J. Biosci. Bio-Technol.* **2016,** *8,* 171–190. https://doi.org/10.14257/ijbsbt.2016.8.1.16.
5. Hussein, A. F.; Arun Kumar, N.; Burbano-Fernandez, M.; Ramirez-Gonzalez, G.; Abdulhay, E.; De Albuquerque, V. H. C. An Automated Remote Cloud-Based Heart Rate Variability Monitoring System. *IEEE Access.* **2018,** *6,* 77055–77064. https://doi.org/10.1109/ACCESS.2018.2831209.
6. El-Nour, A.; Tarique, M.; Foyez, D.; Hassan, L.; Rashed, A. Location Aware Health Monitoring System for Emergency Cases. *Int. J. Sci. Technol. Res.* **2019,** *8,* 1247–1254.
7. Miah, A. In *Continuous Heart Rate and Body Temperature Monitoring System Using Arduino UNO and Android Device*, 2nd Int. Conf. Electr. Inf. Commun. Technol. EICT 2015, 2016, pp 183–188. https://doi.org/10.1109/EICT.2015.7391943.
8. Liu, N. WiGym: Exercise Recognition and Counting Using Wi-Fi Signals. In *Proc. IEEE Int. Conf. Softw. Eng. Serv. Sci. ICSESS* **2018,** *2017-Novem,* 608–611. https://doi.org/10.1109/ICSESS.2017.8342989.
9. Mortazavi, B. J.; Pourhomayoun, M.; Alsheikh, G.; Alshurafa, N.; Lee, S. I.; Sarrafzadeh, M. In *Determining the Single Best Axis for Exercise Repetition Recognition and Counting on Smartwatches*, Proc - 11th Int Conf Wearable Implant Body Sens Networks, BSN 2014; 2014, pp 33–38. https://doi.org/10.1109/BSN.2014.21.
10. Meivel, S.; Charanyameenachi, S.; Supritha, D.; Hamsaveni, S. G. E-Health Real-Time Monitoring System Using IoT Sensor (ESKIN) Methodology; 2019.
11. Qi, J.; Yang, P.; Hanneghan, M.; Tang, S.; Zhou, B. A Hybrid Hierarchical Framework for Gym Physical Activity Recognition and Measurement Using Wearable Sensors. *IEEE Internet Things J.* **2019,** *6,* 1384–1393. https://doi.org/10.1109/JIoT.2018.2846359.
12. Thomas, A.; Gopi, V. P. Accurate Heart Rate Monitoring Method during Physical Exercise from Photoplethysmography Signal. *IEEE Sens. J.* **2019,** *19,* 2298–2304. https://doi.org/10.1109/JSEN.2018.2886001.
13. Kumar, N.; Dahiya, A. K.; Kumar, K.; Tanwar S. In *Application of IoT in Agriculture. 2021 9th International Conference on Reliability, Infocom Technologies and Optimization (Trends and Future Directions) (ICRITO)*; 2021.
14. Kumar, K.; Kumar, N.; Shah, R. Role of IoT to Avoid Spreading of COVID-19. *Int. J. Intell. Netw.* **2020,** *1,* 32–35.
15. Kumar, N.; Kumar, K.; Kumar, A. In *Application of Internet of Things in Image Processing, 2022 IEEE Delhi Section Conference (DELCON)*, 2022; pp 1–5. doi: 10.1109/DELCON54057.2022.9753308.
16. Kumar, K., et al. Dimensions of Internet of Things: Technological Taxonomy Architecture Applications and Open Challenges—A Systematic Review. *Wireless Commun. Mobile Comput.* **2022,** *2022,* 1–23. doi: 10.1155/2022/9148373.

CHAPTER 8

MEDICAL BLOCKCHAIN: A NEW WAY TO REGULATE DATA ON THE "BLOCK"

J. S. RAUTHAN[1], S. K. VERMA[1], SURABHI LINGWAL[1], BANIT NEGI[1], SIDDARTH GHENSAL[2], and Y. S. CHAUHAN[2]

[1]*Department of Computer Science & Engineering, G. B. Pant Institute of Engineering and Technology, Pauri Garhwal, India*

[2] *Department of Computer Applications, G. B. Pant Institute of Engineering and Technology, Pauri Garhwal, India*

ABSTRACT

Electronic health records (EHRs) are compassionate information shared securely with different healthcare providers. Our goal with this chapter was to create a system to store and share EHRs securely using blockchain technology. Blockchain technology is decentralized, immutable, and guarantees non-repudiation of the data. These properties make an opportunity to develop an EHRs system using Blockchain. This chapter has surveyed the relevant related work regarding blockchain technology and the different consensus protocols applied to the Blockchain. The various existing blockchains and platforms that enable the development of decentralized applications are meticulously detailed in the related work section. Health applications that explore topics affiliated with this work, such as using the Blockchain to solve issues with the current healthcare services, were also studied and detailed. This chapter also explained the system model and all the

IoT-Enabled Healthcare Systems: Applications, Benefits, and Challenges
Yatindra Kumar, Shashi Kant Verma, Archana Singh, Krishna Kumar, & Manoj Gupta (Eds.)

implementation considerations to enable the electronic health records of a user to be stored so that the user can access his data and provide it to the different healthcare providers at any time. The proposed architecture, MedicalBlockchain, allows stored data to be viewed, inserted, altered, shared, or deleted by certain entities depending on defined access policies. MedicalBlockchain improves how medical records are shared by ensuring data and owner privacy, service availability, data immutability, and access control management.

8.1 INTRODUCTION

Nowadays, the world is led by information. As a way of dealing with this vast amount of information, people give up their data and preferences to get a personalized view of what exists in the market. All collected data about the population's lifestyle lead brands to adapt their products, and people have become precious targets whose profiles are sold and converted into marketing strategies worldwide.

From this volume of data, several concerns arise as the information being spread by several systems, service availability, and storage security can affect users' privacy.

The concern of data dispersion also affects the health area, which will be the explored area. In Figure 8.1, it is possible to observe the multiple healthcare providers. The concern is our medical data is being spread through all these entities.

It is crucial to understand a health record and why it is essential. A patient's health record is a set of documents that contains information about the health status, personal history, examinations performed, contacts with health services, surgeries, therapeutics, and other data. This set of documents aims to improve healthcare efficiency and effectiveness, so the health record must be available when necessary. Other purposes need this information, such as clinical research and health services management.[1]

The Electronic Health Record (EHR) is the subset of Health Record or Medical Record that is digitally stored information for simplifying shared care, training, and research. The EHR is a tool to help the continuity of the quality, access, and efficiency of healthcare delivery.[2]

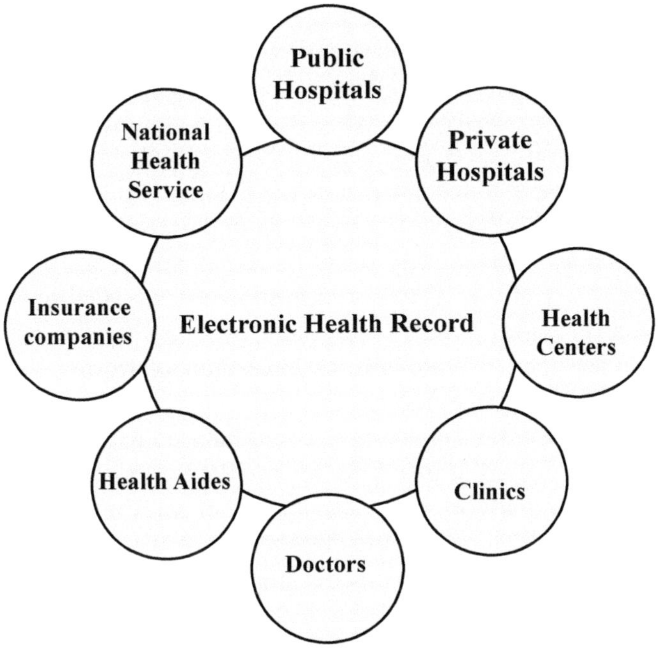

FIGURE 8.1 Healthcare providers.

8.1.1 MOTIVATION

The EHRs are stored in each healthcare provider in centralized databases; the Governance consequently exists as a central entity that could be considered a single point of failure. Another problem in healthcare services is the fragmentation of health records.[3] Electronic Health Records are fragmented across hospitals, health centers, doctors, insurance companies, and wearable devices. Patients cannot precisely describe what was said by one doctor, and the lack of information-sharing mechanisms between medical entities can lead to inaccurate diagnoses in different appointments. It is essential to have a system that allows all the participants to view, insert, modify, share, and delete EHRs to increase the probability of an accurate diagnosis and more effective treatments.

However, data is a sensitive matter. This way, the storage mechanisms used should have precautions related to security, integrity, and availability of the data and different access control levels to manage what entities get to consult what type of data. Another issue to be solved is service availability and data immutability.

It was decided to study and create a user application that uses Blockchain technology as a Back-end service to solve these problems. This makes a unified solution for all the entities related to the Healthcare system and discards the necessity of relying trust on a central authority. Because Blockchain technology is immutable, the data will be preserved, and it will also be stored in a decentralized way.

In Portugal, an attempt was made to alleviate the challenge of data fragmentation by empowering users to access and share their electronic health records (EHRs). The Portuguese National Health Service[4] has a portal that provides access to the data collected by the public service and private services. Private entities have their platforms, as happens in Hospital da Luz[5] and Hospital Lusiadas.[6]

To solve the system's challenges could be used a centralized database. Still, these types of solutions store all the data on a single database and highly depend on network connectivity. Other relevant issues are related to availability, scalability, and business model. A distributed database was not chosen to solve this problem because, in this type of database, it is assumed that all the data should be distributed among several computers, and all nodes trust each other and will not misbehave or leak data.[7]

Given the system's challenges, such as access control, information preservation, data authentication, and non-repudiation mechanisms, a Blockchain is a possible approach that has these features and solves the main problem of data fragmentation.

Blockchain was introduced in the paper Bitcoin: A Peer-to-Peer Electronic Cash System.[8]

This technology was created to solve the double-spending problem so that transactions could be carried out without a central system dealing with such transfers in banking institutions. By definition, according to Ref. [9], a blockchain is a decentralized, transparent ledger for transaction records—an infrastructure that is shared by all network nodes, updated by miners, monitored by everyone, and owned and controlled by no one. To support the solution, it uses a permissioned blockchain to know who the entities involved in the network are.

8.1.2 OBJECTIVES

The main goal of this chapter was to enable the electronic health records created by different healthcare providers to be stored, so the user can access his data and provide it to the various healthcare providers at any time. Depending on defined access policies, it is allowed to store data to be viewed, inserted, altered, shared, or deleted by certain entities.

Blockchain technology solves the data fragmentation problem and helps preserve all the data collected about the patient. The information stored in the chain cannot be altered or deleted, but it can be modified from a user's perspective by adding new information to the chain. It provides an efficient exchange of health records between entities without compromising data privacy and respecting the owner's privacy.

By law, the user is the owner of his personal information. This way, the user has the right to access his clinical process, to define a set of policies that state who can access his data and even if his medical history has to be preserved.

8.2 THEORETICAL BACKGROUND

This section will detail the Blockchain technology, consensus protocols, and the main applications that use Blockchain technology, Ethereum, and Smart Contracts. It will describe technologies that can help accomplish the chapter's goals and solutions that solve a parcel of the problems in EHRs. At the end of this section, it is detailed the legal framework regarding sensitive data.

8.2.1 BLOCKCHAIN

The technology known as the Blockchain was introduced by Satoshi Nakamoto[8] in 2008 to solve the problem of Double Spending in digital currency. The author/s introduced the concept of Bitcoin, which is the digital currency that uses Blockchain as technology.

Blockchain is a decentralized peer-to-peer (P2P) system that is encrypted to ensure that only authorized users may access the data. Blockchain technology is based on the peer-to-peer (P2P) protocol used by BitTorrent.[10] Blockchain technology combines peer-to-peer networks with a distributed consensus mechanism to tackle the synchronization problem of traditional distributed databases.[11] Blockchain and Alternative Payment Models,[12] Blockchain can be thought of as a protocol for resolving the Byzantine Generals' Problem,[13] a well-known computer science problem involving reaching consensus in a decentralized system with defects. This technology is composed of six key attributes:[11]

Decentralized: Blockchain is a P2P system, so it does not need a central server, all machines are equal, and each machine provides a piece of its resources. In a P2P system, potentially, the higher the number of machines, the faster communications will be and consequently the service performance.

a. **Transparent:** The stored data in the Blockchain can be seen by all stakeholders.

b. **Open Source:** There are no restrictions on the creation of blockchain-based applications.

c. **Autonomous:** Due to the consensus protocol, any node can safely transfer or update data by trusting the system and not a single person.

d. **Immutable:** Unless an entity has control over 51% of the network nodes, it is impossible to delete or modify any information contained in the Blockchain.

e. **Anonymity:** It is only necessary to know the person's blockchain address to execute a transaction. This address cannot be linked to the owner's actual IP address.

It is considered the existence of three types of blockchains:[11,14,15]

a. **Public Blockchain:** Anyone can read or perform transactions and expect them to be verified. Participation in the consensus process is open to all nodes. A consensus process will decide which blocks are inserted and the current state. The decentralized nature of a public blockchain is well-known in the technology community. Public blockchains like Bitcoin and Ethereum are good examples.

c. **Consortium Blockchain:** The ability to read blocks can be granted to anyone, or it might be reserved for a small group of people. The network configuration specifies which nodes are involved in the consensus process, and only those nodes can participate. The majority of consortium blockchains are regarded as partial-decentralized. A consortium blockchain is Hyperledger.

d. **Fully Private Blockchain:** In contrast to read permissions, write permissions can only be granted to a single person or organization.

Public blockchains are also known as permissionless, and consortium and private blockchains are known as permissioned blockchains. Permissioned blockchains are faster, easier to implement, and energy-efficient.[16] These types of blockchains can support consensus processes other than Proof-of-Work[10] or Proof-of-Stake,[17] which consume a high quantity of energy. Also, the throughput of these types of consensus processes could be lower than the consensus adopted in permissioned blockchain implementations, such as implementing a Byzantine Fault Tolerance consensus algorithm.[18]

The Blockchain is a chain of blocks, as shown in Figure 8.2. A block is always dependent on the previous block. A block is created by the nodes that participate in the consensus protocol. Each block is composed of two parts, according to Ref. [19], the block header and the block body, as illustrated in Figure 8.3.

The following are standard components of a block header:

a. The *block version* identifies which set of block validation criteria should be used.
b. *Merkle tree* root hash is the hash value of all the transactions in the block.
c. Using the universal time since January 1, 1970, the *timestamp* is the current time in seconds.
d. *nBits* is the minimum number of bits required for a block hash to be valid.
e. The *Nonce* is a four-byte field that is generally initialized to 0 and rises with each hash calculation.
f. The *parent block hash* is a 32-byte hash value that identifies the previous block in the Blockchain.

A block's body comprises the total numbers of transactions and the actual transactions themselves, depicted in Figure 8.3. In general, the blockchain protocol works as follows:[11]

a. The node creates new information and broadcasts it to all network nodes.

b. A node creates a new block when it receives correct information and has enough info stored to create a block with a minimum number of transactions.
c. The nodes execute the consensus algorithm on the block. If consensus is successful, the block will be added to the chain.

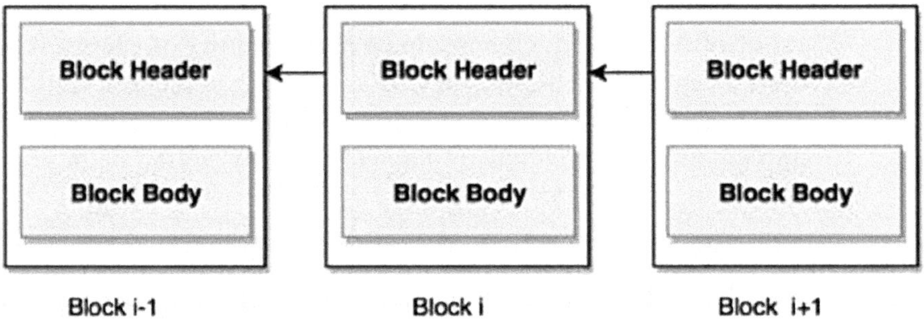

FIGURE 8.2 Blockchain.

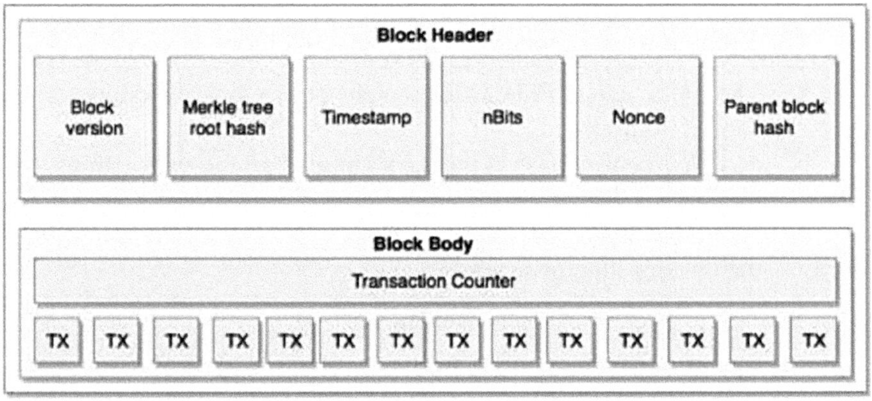

FIGURE 8.3 Block structure.

Due to the security requirements, a Consortium Blockchain is the type of Blockchain that could help achieve the chapter's goal. The consortium blockchain allows the definition of access permission types that restrict access to the medical records of each user.

We further conclude that transactions are digitally signed, leading to non-repudiation of the data. All entities keep a copy of the transactions verified, ensuring that it has not been modified.[20] It will also be possible to solve the problem of data preservation because the Blockchain is immutable.

The two most well-known applications of the blockchain data structure are Bitcoin and Ethereum.

8.2.2 CONSENSUS

Consensus algorithms aim to make a uniform decision.[16] A consensus protocol ensures *agreement*; all honest nodes will agree on which transactions are selected, *termination* in which all honest nodes eventually learn the chosen transaction, and *validity*, which guarantees that the selected transactions were actually proposed by some node.[21]

There are many consensus protocols that can be applied to the Blockchain, such as Proof-of-Work (PoW), Proof-of-Stake (PoS), Proof-of-Existence (PoE), Proof-of-Activity (PoW and PoS hybrid), Proof-of-Elapsed Time (PoET), Proof-of-Importance (PoI), Delegated-Proof-of-Stake (DPoS), Practical Byzantine Fault Tolerance, Ripple, Tendermint, among many others.[16]

8.2.2.1 PROOF-OF-WORK (POW)

The consensus protocol used in Bitcoin and Ethereum is Proof of Work (PoW). According to the creator of Bitcoin,[8] the goal of PoW is to find a value that when the block header is hashed with SHA-256, it begins with a certain number of zero bits. This protocol is based on a similar system to Adam Back's Hashcash.[22]

It is crucial to understand the target to understand the Proof-of-Work. The "*Target*" is a 256-bit number shared by all Bitcoin clients and is set, so the mining time limit is 10 min. According to Mastering Bitcoin,[10] the target value will be modified every 2016 block[11] using the following equation:

$$NewTarget = OldTarget * (ActualTimeofLast2016Blocks/20160minutes) \quad (8.1)$$

The nodes that execute this consensus protocol are called miners. Miners will construct a block that contains transactions. Then, the miner will calculate the hash of the block's header and verify if it is lower than the current target. If the hash is not lower than the target, the miner will increment the Nonce by one and repeat the process.[10] When a miner can get the desired value, it will broadcast the Nonce to verify other miners in the network. If the number of mining nodes, defined in the consensus rules by the developers, validates the Nonce, the consensus is achieved, and a new block will be inserted into the chain.[19]

If it is possible to find a Nonce such that the header's hash is smaller than the target in less than 10 min, the difficulty[1] will increase; otherwise, it decreases.

PoW is essentially a one-CPU-one-vote protocol because SHA-256 is an entirely unpredictable pseudo-random function. The only way to create a valid block is by brute-force testing different values for the Nonce.[23]

This consensus mechanism will prevent Sybil attacks[2] because creating a block requires considerable processing, resulting in a waste of resources. Proof-of-Work can be performed by any node in the network, as long as it has enough computational power.[23]

8.2.2.2 PROOF-OF-STAKE (POS)

Compared to Proof-of-Work, the algorithm Proof-of-Stake (PoS) is an energy-saving solution.[25] PoS replaces PoW's competition by offering the authority to update the Blockchain to a randomly selected stakeholder.[26] PoS offers an explicit monetary reward to update the Blockchain, but it does not impose any direct cost upon agents to gain the authority to edit the Blockchain.[26] For the PoS algorithm, there are different types of implementations like Nxt,[27] BlockCoin,[28] or PPCoin[17] that are explained in more detail.

The PPCoin algorithm introduced in 2012 in the article PPCoin: *Peer-to-Peer Crypto-Currency with Proof-of-Stake*[17] is a hybrid solution between PoS and PoW. This article introduced the concept of the coin age, which the authors define as "currency amount times holding period."[17] For example, if Bob received 10 coins from Alice and did not spend them for 90 days, Bob has 900 coin-days of coin age.

PoS blocks contain a new type of transaction called coinstake. In this type of transaction, the owner will consume his coin age and gain privileges to generate a new block and mint for proof-of-Stake.

The coinstake transaction is composed of the Kernel input and the Stake input. The Stake input is the reward added by the nodes that will mine this block. The Kernel input has to follow a specific hash target protocol to create the block similarly to proof-of-Work blocks.

The hash target in PoS is a coin-day consumed, whereas, in Bitcoin, it is a fixed target applied to every node. Therefore, in PoS, finding the hash is dependent on the coin age consumed; if users have more coin age, the easier it is to discover the hash target.

In this new type of consensus, it is possible to reduce the possibility of malicious attacks. This is given the fact that to carry out an attack is expensive, and it will reduce the incentives for the attacker since he would have to own the majority bitcoin to mint the block and consume his coin age.[11]

The consensus protocol used by Ethereum is PoW, but the creator of Ethereum declared that it would be modified to a hybrid version of PoW and PoS, which will be called Casper.[29]

8.2.2.3 PRACTICAL BYZANTINE FAULT TOLERANCE (PBFT)

Practical Byzantine Fault Tolerance (PBFT) is a replication algorithm that supports Byzantine faults. This protocol was created to solve Byzantine Generals Problems.[30] PBFT is only used in a permissioned blockchain where all network stakeholders are known.

The components of this protocol are the primary node and replicas. The total number of replicas will be equal to $n = 3f + 1$, where n is the total number of replicas and f is the number of failures the system can handle. It is assumed that having more than $(n - 1) = 3$ faulty replicas is impossible to ensure safety and liveness proprieties.

In Figure 8.4 are illustrated the three phases of this algorithm: pre-prepare, prepare, and commit. At each stage, a node vote for a value until it reaches a byzantine quorum; it is only moved to the next phase if each node has received the answer of (2/3) of the nodes.[19]

Cryptographic methods such as digital signatures, message authentication codes, and hash functions prevent this algorithm from spoofing attacks, replay attacks, and detecting corrupted messages. An access control mechanism was created for customers to authenticate themselves, which prohibits access if they are trying to request a resource that they do not have permission to access.

In general, the algorithm works as follows:[30]

- The primary node receives digitally signed requests containing a timestamp from the client and will multicast to the replicas;

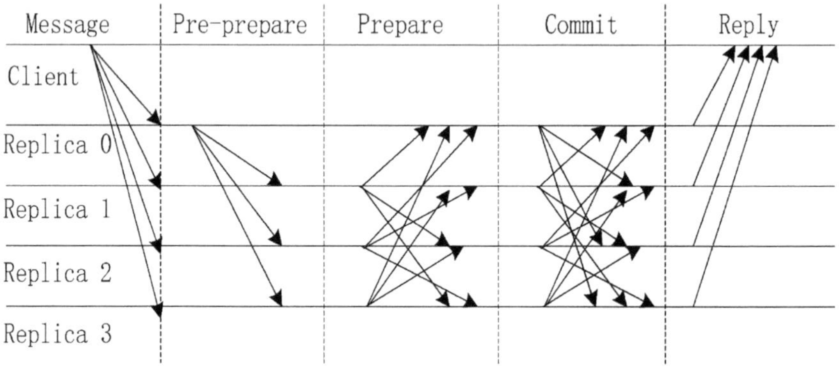

FIGURE 8.4 Practical Byzantine Fault Tolerance (PBFT).

- Replicas process the query and deliver the client a digitally signed answer;
- The client awaits f + 1 responses from several replicas. If the digital signatures are valid, timestamps and the results of the request are equal; it becomes the result of the request made.

If the replies do not arrive in time, the client broadcasts the request to replicas. If the request was already processed, the replicas resend the response to the client and send the request to the primary node; replicas save the last reply forwarded to each client. If the primary node does not multicast the request to the replicas, it will eventually become suspected and may occur a view change, which means that the primary node was modified.

PBFT could be adapted to a blockchain solution, as examples of these adaptions are: BFT-SMaRt[31] and Tendermint.

8.2.2.4 RIPPLE

In the Ripple consensus protocol,[32] nodes are divided into two types, the Server that participates in the consensus process and the client that will transfer funds.

This consensus algorithm is composed of six components:[19,32]

- The S*erver* executes the Ripple Server software and will take part in the consensus process;

- The *Ledger* keeps the amount of money in each account and is updated when a transaction successfully completes the consensus;
- *Last-Closed Ledger* is the Ledger where the consensus process is successfully applied;
- *Open ledger* store the transactions while the consensus process is taking place;
- *Unique Node List (UNL)* is a list of the servers to communicate to complete the consensus process;
- *Proposers* are the only ones that can propose a value during the consensus process.

This consensus protocol works in rounds and each round:[19,32]

a. Each server forms a public list with all valid transactions that have not gone through the consensus process, corresponding to the "candidate set";
b. Each server will join all the "candidate sets" in the UNL and vote to send their opinion about the transactions;
c. Transactions that earn more than the minimum percentage of positive votes can go to the next round. If this does not happen, they will be discarded or included in the candidate set again;
d. In the last round, it is mandatory to have at least 80% of the servers from UNL agree on the transaction, which means that at least 80% of validating servers from the UNL should have the same set of transactions;
e. If all transactions meet these requirements, the ledger will be closed and become the last-closed ledger.

8.2.2.5 TENDERMINT

Terdermint[33] is a Byzantine consensus algorithm. The nodes are connected using a peer-to-peer architecture, and the information is transmitted through gossip. The nodes are designated as validators because these are the ones who propose the blocks and vote. This algorithm is divided into four steps:[19,33]

a. *Propose*: The chosen proposer for that round will broadcast a block to the neighbors by gossip.

b. *Prevote*: Each validator must decide on the block's validity and broadcast the decision. In case the block is invalid or does not arrive on time, its prevote will be nil.
c. *Precommit*: If a node receives more than 2/3 of prevotes, it does a broadcast of the precommit of the block; otherwise, it sends NIL. If this node receives 2/3 of precommit, it commits.
d. *Commit*: If the node validates the block, broadcasts the commit, and receives 2/3 of commits, it accepts the block.

8.2.2.6 SUMMARY OF CONSENSUS ALGORITHMS

Table 8.1 demonstrates that in a permissioned blockchain, the main consensus algorithms used are the Practical Byzantine Fault Tolerance and the Tendermint algorithms. It is illustrated that only the PBFT, Ripple, and Tendermint are consensus algorithms that are energy savers. The third column of this table represents the percentage of faults tolerated by the algorithms to finalize consensus successfully.

TABLE 8.1 Effect of pH on the Adsorption Capacity of Modified Bentonite.

Consensus protocol	Node identity management	Energy saving	Tolerated power of adversary	Example
PoW	Open	No	<51% computing power	Bitcoin/Etheruem
PoS	Open	Partial	<51% stake	Peercoin
PBFT	Permissioned	Yes	<33.3% faulty replicas	—
Ripple	Open	Yes	<20% faulty nodes in UNL	Ripple
Tendermint	Permissioned	Yes	<33.3% byzantine voting power	Tendermint

8.2.3 ETHEREUM AND SMART CONTRACTS

Ethereum's purpose was to design a new protocol for creating decentralized apps. Blockchain with a built-in Turing-complete programming language produces an additional protocol that permits anyone to write a smart contract and decentralized applications to specify their ownership and transaction type laws.

Nick Szabo was the first to propose the idea of smart contracts in 1994. A smart contract is described by the author of the article[25] as a computerized transaction protocol that performs the provisions of a contract. Smart contracts are cryptographic "boxes" that store value and may only be unlocked if particular criteria are satisfied, according to Ethereum's developer.[23] Because smart contracts have incorporated Turing-completeness, value-awareness, blockchain-awareness, and state, they have far more power than Bitcoin scripting.

The Ethereum Virtual Machine (EVM) is Ethereum's most significant breakthrough. EVM is a decentralized computer with millions of "accounts" or objects. It also presented the idea of the message and reformulated the notion of transaction.

Accounts have four fields:

- *Nonce* represents the counter to ensure that a transaction is handled only once;
- *Ether balance* (digital currency of Ethereum) of the account;
- *Account contract code*, if required;
- *Account storage* (empty by default).

The accounts are responsible for maintaining internal databases, executing code, and communicating with one another. There are two types of accounts, according to Ethereum's inventor: externally owned accounts and contract accounts. The private key controls externally owned accounts, the contract field is empty, and a message can be sent by generating and digitally signing a transaction. The contract code controls contract accounts; whenever this type of account obtains a message, the contract code is performed, permitting reads and writes on internal storage, sending additional messages, and creating contracts.

The EVM[35] executes the smart contract, built-in Solidity, a contract-oriented high-level language with a syntax comparable to JavaScript.[36] These contracts are Ethereum's execution environment's independent agents. When initiated by a message or a transaction, smart contracts are performed. Smart contracts have access to the ether balance and the saved data in the account to guarantee compliance with the persistent variables.[23]

In Ethereum, a transaction is a signed data packet containing the message to be forwarded to an externally owned account.

The transaction is composed by:[23]

- The receiver of the message;
- Identification of the sender through the digital signature;
- Quantity of ether to be transferred;
- Optional data field that by default has no defined function, but each EVM has an opcode with which the contract code can retrieve this field;
- STARTGAS represents the most significant number of computational steps the transaction is allowed to perform;
- GASPRICE represents the rate that the sender has to pay per computational step;

The last two fields define the transaction's cost and are critical for implementing an anti-denial-of-service architecture. Each transaction must specify a limit on the number of computational steps required to execute code to avoid infinite loops or computational waste.

When a transaction is added to the block or when the block is validated, the contract will be executed. Each contract has a unique address that identifies it. All nodes perform the contract using their EVMs and agree on the contract's output, and the new state is obtained in the consensus.[23,37]

Messages are similar to transactions, except they are generated by other contracts, not by a third party. The message is constructed and executed when a contract performs the CALL opcode system.

The components of a message are:[23]

- Message Sender (implicitly);
- To whom the message will be sent;
- Amount of ether transferred with the message;
- Optional data field;
- STARTGAS value.

According to Ref. [23], the block validation algorithm in Ethereum has to:

a. Ascertain the existence and validity of the preceding block;
b. The block's timestamp is more than the preceding block's but less than 15 minutes in the future;
c. Ascertain the validity of the block number, difficulty, transaction root, uncle root, and gas limit;
d. Verify if the proof of work has been done with success;

e. Save the most recent state in the block;
f. Determine whether the Merkle tree root of the most recent state corresponds to the state specified in the block header.

The primary distinction between Ethereum and Bitcoin is the smart contract and block component architectures. Each block in Ethereum contains a replica of the list of transactions, the most recent state, the block number, and the consensus protocol's difficulty.

8.2.4 PLATFORMS

Several platforms have already explored Blockchain as a technology. The platforms studied in more detail are Quorum, Chain Core, Corda, and Hyperledger platforms. Microsoft Azure supports Chain Core, Hyperledger Fabric, Corda, and among others, Blockchain as a Service (BaaS).

8.2.4.1 HYPERLEDGER

Hyperledger is a Linux Foundation-hosted open source project. Hyperledger's objective is to make an open-source technology that enables the building of distributed ledgers to facilitate corporate transactions.

Hyperledger is composed of four frameworks:

a. Hyperledger Burrow;[42]
b. Hyperledger Iroha;[43]
c. Hyperledger Sawtooth;[44]
d. Hyperledger Fabric.[45]

Hyperledger Burrow[42] was created by Monax[46] and co-sponsored by Intel. Burrow aims to create a secure framework and ensure data privacy. This framework executes Ethereum smart contracts using a permissioned virtual machine, an adaptation of the EVM. To develop the smart contract, Monax provides the *monax chains*. The consensus protocol used in this framework is Tendermint.

Hyperledger Iroha[43] authors are developing a blockchain with C++ emphasizing mobile applications and have created a chain-based Byzantine Fault Tolerant consensus algorithm described as *Yet Another Consensus*. Iroha is designed to build and manage custom complex assets, user accounts, access control policies for transactions execution and queries, and validation of business rules for transactions and queries. Blockchain-based applications could be written in Python, Java, JavaScript, C++, and Android and iOS mobile platforms.

Hyperledger Sawtooth[44] is an enterprise blockchain project. In this framework, the concept of *Transaction families* was created that correspond to the smart contracts in Ethereum. The transaction families are used to define the business logic. Sawtooth has two implementations that are used as consensus algorithms. The first one is Dev_mode; this algorithm is explained in the Sawtooth documentation[44] as "*a simpler random leader technique for developers and test networks with minimal crash fault tolerance requirements.*" The second one is Proof of Elapsed Time (PoET),[47] which is one of the solutions for the Byzantine Generals Problem. PoET uses a trustworthy execution setting to increase the effectiveness compared to Proof-of-Work. The transaction families and the consensus algorithm will help decide the state of the data in the Blockchain.

IBM and Digital Asset initially contributed Hyperledger Fabric.[45] This framework is a permissioned blockchain. Hyperledger Fabric was designed to be modular and allow pluggable and custom components such as consensus and membership services. The Membership Service Provider (MSP) abstracts the cryptographic mechanisms and protocols behind issuing and validating certificates and authentication for participants within the blockchain network. Only participants who are part of the same MSP can communicate. Hyperledger Fabric created the concept of Chaincode, which corresponds to smart contracts. The chain code will be responsible for the business logic, and solely this code can read or manipulate the data on the chain. The four main components of this framework are:

a. *Fabric CA* generates certificates for the users;
b. *Peer* stores the Blockchain;
c. *Endorsing Peer* receives and simulates the transaction and executes the chain code;
d. *Ordering Service* guarantees the order of the transactions, creates the block, and sends it to the peer.

Consensus is attained when the order and outcome of each block's transactions are validated. Solo Order and Apache Kafka are the techniques utilized for ordering.[48]

In this framework exists the concept of Channels. A Channel is a private blockchain overlay that allows data isolation and confidentiality.[49] Peers can be part of one or more channels. Each peer has its own identity when they join the channel; this identity is given by the MSP and will authenticate each peer to its channel peers and services. Each channel has associated at least one chain code responsible for that channel's ledger.

Besides these frameworks, Hyperledger has tools to make blockchain access and development more accessible and practical. Currently, three tools are available:

a. Hyperledger Cello aims to reduce the effort needed to create, manage, and terminate the Blockchain.
b. Hyperledger Composer is a set of tools that makes blockchain business network, development of smart contracts, and the deployment beyond the distributed ledger.
c. Hyperledger Explorer creates a user-friendly Web application to view, invoke, deploy, and query blocks, transactions, and other information stored in the ledger.

8.2.4.2 SUMMARY OF THE STUDIED PLATFORMS

Table 8.2 presents a summary of the different platforms studied. All the platforms are permissioned blockchains. The second column explains how transactions are processed and the last column the consensus mechanism used.

8.2.5 HEALTH APPLICATIONS

Many authors and companies created applications to apply blockchain as the technology used in healthcare systems. In this section, it will be presented four examples of architectures that used blockchain to create a healthcare system.

8.2.5.1 MEDREC

MedRec[50] tried to solve the following problems of a users' Electronic Medical Records (EMR):

- Fragmented and slow access to medical data;

TABLE 8.2 Platforms.

Platforms name	Transaction flow	Consensus mechanism
Quorum	Private transactions propagated in a subset of nodes, while public transactions are network-wide	Raft-based or Istanbul BFT
Chain Core	Bitcoin-like	Federated consensus protocol
Corda	Bitcoin-like	Validity and Uniqueness consensus
Hyperledger Burrow	Transactions are sent to the consensus mechanism, blocks are then propagated in the network.	Tendermint
Hyperledger Iroha	Transactions are proposed to the peers, then is applied the consensus mechanism, and finally, the block is sent to the peers.	Yet Another Consensus
Hyperledger Sawtooth	Bitcoin-like	PoET or Dev mode
Hyperledger Fabric	Transactions are proposed to the endorsing service, then is applied the consensus mechanism, and finally, the block is sent to the other peers.	Apache Kafka or Solo Order

- Interoperability of the systems;
- Patients are not fully informed of their medical history and any modifications to it.
- Improved data quality and quantity for medical research.

The authors chose a public blockchain, specifically the Ethereum blockchain. The authors choose Ethereum blockchain to use smart contracts to contain metadata on record ownership, permission, and data integrity. EMRs are stored in external databases, the blockchain only contains pointers to these databases. The blockchain contains a hash of the saved data to ensure data integrity. When an entity adds data about a patient it sends a notification to the patient and the patient can verify it before accepting or rejecting. He/she will be able to share his/her data with other entities. To verify the participant's entity is used the public key and a DNS-like mapping between the Ethereum address and the client's social security number or name.

In order to define the relations between the entities and users, three types of contracts were created:

- **Registrar Contract (RC)** is used to map the participant identity with the Ethereum address, which is equivalent to the public key, and have the Summary Contract;
- **Patient–Provider Relationship Contract (PPR)** has the permissions to access data;
- **Summary Contract (SC)** has references to the PPRs, representing all interactions made with the system nodes.

In Figure 8.5 it can be seen that the architecture of this system has four components:

a. **Backend Library** aims to abstract communications with the blockchain;
b. **Ethereum Client** will create peer-to-peer connections on the network, encrypt, and send transactions. Ethereum Client also stores a local copy of the blockchain;
c. **Database Gatekeeper** implements an off-chain, it has a pointer to a local SQLite database and it is possible to get access to it depending on the permissions stored in the blockchain;
d. **EMR Manager** is composed of the above components. This component will be used for EMR management and for user interface applications. Users will receive notifications when medical data is updated and can share and retrieve data.

The authors of this architecture used Ethereum's consensus protocol, Proof of Work. Figure 8.5 illustrates how the system will operate if a provider inserts medical records for a new patient.

8.2.5.2 ONCOLOGY-SPECIFIC DATA MANAGEMENT

The framework proposed in Ref. [51], was developed to manage oncology data. This platform was created to allow users to define access control policies of their data. Another important feature of this platform is to improve the sharing of this data among the various entities. To create this framework a permissioned blockchain was chosen for the following reasons:

- Importance of knowing the user's identity;
- Health records are sensitive data, so monitoring the communications between the patient and the entities can expose sensitive data, violating the privacy of the user;
- quick response of the system when doing any procedure is crucial for the user;
- If an entity has to pay to make changes to the user's health records, it could limit the system's usability.

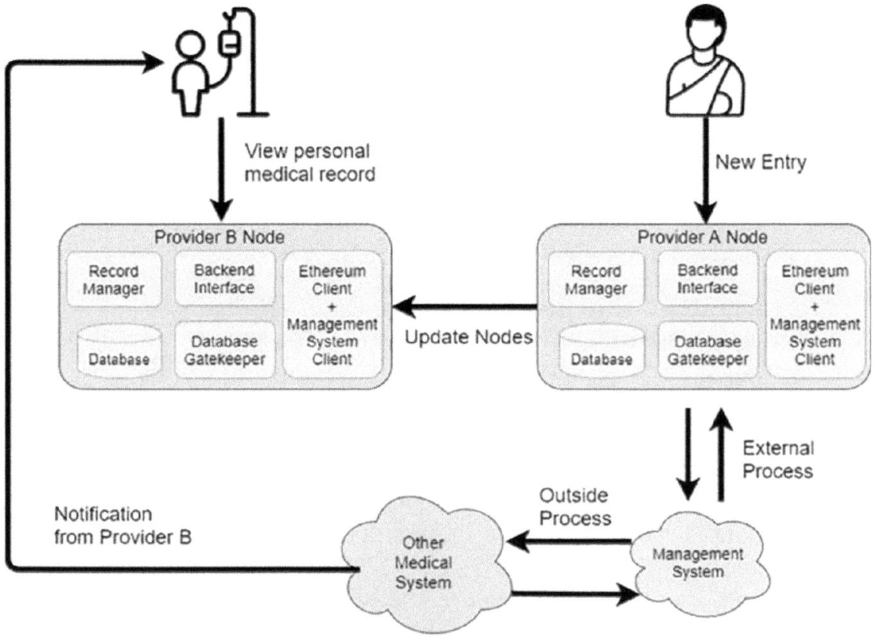

FIGURE 8.5 MedRec's use case: add a record for a new patient.

***Source*:** Reprinted with permission from Ref. [87] https://creativecommons.org/licenses/by/3.0/

This framework is composed of four components:

a. Membership service records the roles of system actors. The roles define the blockchain functionalities that are available.
b. Databases store health records off-chain.
c. Nodes address the consensus protocol.
d. APIs for different types of users.

User data is collected in two different locations. A local database is placed in the hospital and a cloud where data is organized according to sensitivity degree and encrypted with the user's symmetric key. An entity can retrieve data in the cloud depending on the access control policies defined by the client.

To implement the blockchain technology it is used Hyperledger platform. All nodes will be validating peers and will receive all transactions performed by users through role-based APIs. The consensus protocol used was Practical Byzantine Fault Tolerance.

8.2.5.3 HEALTHCHAIN

The Health Insurance Portability and Accountability Act of 1996 (HIPAA) was utilized in the United States to guide the development of the Healthchain[52] protocol. The term "Protected Health Information" was defined in HIPPA (PHI). Individually identifiable health information, such as demographics and genetics, is protected under PHI, which establishes strict guidelines for its handling. The HIPAA privacy rule establishes guidelines for safeguarding the personal health information (PHI) of persons while also guaranteeing the patients' sovereignty over such data.

Using this method, authorized individuals can access, evaluate, and edit a record that has been agreed upon. An initial version of PHI will be inserted into the blockchain by the patient. Smart contracts ensure that only the original version of a transaction is inserted into the blockchain by the user system.

Healthchain uses three components in the architecture:

a. Private Blockchain on IBM Blockchain;
b. Deployed on Bluemix, IBM cloud;[53]
c. Web page, where the NodeJs server is responsible for the communication with Hyperledger fabric and the chaincode.

Data privacy, scalability, and security are all assured by Hyperledger Fabric. Using smart contracts, network nodes can define their own rights and privileges. The Practical Byzantine Fault Tolerance consensus mechanism is utilized by Healthchain as a standard. Until consensus is reached, transactions in this system are not complete. Nodes with the appropriate permissions can only see updates to PHI that have been made.

8.2.5.4 MEDICALCHAIN

When it comes to electronic health records, MedicalChain[3] employs blockchain technology. Health records are stored on the blockchain, and a single copy of each user's data is maintained. Hyperledger Fabric implements a private blockchain and enables granular control over who has access to which data, both of which are essential to maintaining data privacy. Medical records are stored on the platform. They'll scan and upload the physical papers to their cloud storage account. The electronic health record (EHR) will be standardized, ciphered, and transferred to a cloud.

MedicalChain's designers take the following steps to meet their objectives:

a. Wearable devices, doctor notes, scans, and medication dispensing all generate data.
b. Using a cloud storage service, data is encrypted before being delivered to the cloud. A unique identifier for cloud-stored data will be kept on the patient blockchain;
c. In order to access encrypted data, the blockchain ID is utilized when data is requested.
d. The encrypted data is decrypted and shown on the device or application concerned.

MedicalChain's functions include:

- control over one's own personal information is given to the user in this application.
- gives you a record of when and what kind of data you've accessed;
- Data can be accessed and verified by insurance providers. This verification will guarantee that the treatment the patient is receiving is in accordance with what they expected.

8.2.5.5 SUMMARY OF THE HEALTH APPLICATIONS

Table 8.3 summarizes the findings of the comparative analysis between various health applications. Blockchains that are permissioned are the most common form utilized to solve challenges in healthcare services. In order to develop this form of blockchain, Hyperledger Fabric was most often used.

To deploy the blockchain, MedRec's design relies on Ethereum and uses the PoW consensus mechanism, both of which are resource-intensive. Oncology's framework was built to contain solely Oncological information. HealthChain's architecture is equivalent to Oncology's, but it keeps all forms of medical data but, like MedicalChain, it also saves the data in Cloud, which decreases system availability.

TABLE 8.3 Summary of the Health Applications.

Type of applications	Type of blockchain	Consensus	Data	Platform
MedRec	Permissionless	PoW	Database	Ethereum
Oncology	Permissioned	PBFT	Database and Cloud	Hyperledger Fabric
HealthChain	Permissioned	PBFT	Cloud	Hyperledger Fabric
MedicalChain	Permissioned	—	Cloud	Hyperledger Fabric

8.3 SYSTEM MODEL

Using blockchain technology, this section aims to establish an EHR storage and sharing system that is both secure and scalable. It is critical to protect the user's identity and data privacy when it comes to healthcare records. In order to have access to a patient's complete medical history, it is essential to ensure that the data is preserved. Access control measures and the ability to identify who accessed the data are critical in this type of data. As the last step, we must ensure that the system is always available.

8.3.1 SYSTEM COMPONENTS

In this section, you will learn how to build a healthcare application that employs Blockchain technology. "MedicalBlockchain" is the name given to this application software.

An essential decision in this project was to use a permissioned blockchain implementation. Because of its compatibility with both Blockchain and health-related applications, Hyperledger Fabric was selected.

Figure 8.6 shows the three most important high-level components of the system.

The user application was constructed as a proof-of-concept for the suggested architecture, although it was not the primary goal of this project. To establish that blockchain technology can be used to create a healthcare system, the developers created an application that could be utilized by real people.

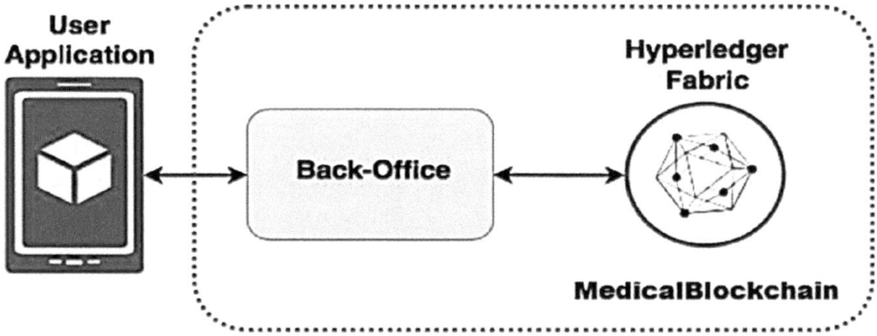

FIGURE 8.6 MedicalBlockchain high-level architecture.

To access the Hyperledger Fabric network, requests are sent from the user application to the Back-office.

Specific criteria were determined based on the functions and use-cases. Table 8.4 lists various roles, each with its own set of predefined access controls.

TABLE 8.4 Roles and Access Controls.

Roles/access controls	View	Insert	Modify	Share	Delete
Patient	Yes	Yes	Yes[1]	Yes	Yes[1]
Doctor	Yes	Yes	—	Yes[2]	—
Health aides	Yes	Yes	—	—	—
Insurance companies	Yes	—	—	—	—
Emergency services	Yes	Yes	—	Yes[2]	—

[1]Medical record inserted by the patient.

[2]With patient approval and only with healthcare providers.

Individuals have access to and control over their own medical records. Patients can revoke access permissions at any moment. Doctors have accessibility to the same features as patients, with the exception of the ability to edit or delete a patient's medical record. Like nurses, health aides can only access and add new medical records. Insurers are only permitted to see medical records that patients have given them permission to see. All emergency service healthcare providers can only share patient data with a healthcare professional if the patient has given their permission. When a patient registers on the system, he or she must give permission for emergency services to communicate his or her medical information with a healthcare professional. This permission can be altered at any moment.

All the users have the option of deleting your account at any moment and erasing all of your history, which means that all of your data will no longer be available.

The Back-office is able to register and authenticate users in order to enable the definition of different user kinds. Authenticating a user requires the user's username and password, respectively. Passwords are generated during the registration process and are unique to each individual user. It is the patient's number that serves as the patient's unique identification. The medical licence number serves as a unique identifier for doctors, health workers, and emergency services. It will be referred to as the insurances' code by insurance companies. In a real-world scenario, a specialized authority should verify the unique identifier.

8.3.1.1 FUNCTIONALITY

The following characteristics were added to the system in order to meet its objectives:

1. Administrator:
   - User registration;
   - User Login;
   - System monitoring.

2. Patient:
   - Insert/View patient's personal information;
   - Insert/View/Share/Medical Record;
   - Modify patient's personal information;
   - Delete Medical Record created by the patient;
   - Share Medical Records on Emergency State;

- Withdraw access privileges;
- Delete account;
- Notify the user when an operation is executed on his/her Medical Record.

3. Medical Staff:
   - Insert/View Medical Record;
   - Share Medical Record with other medical staff;
4. Emergency Service:
   - Insert/ View Emergency Medical Record;
   - Share Emergency Medical Record with other medical staff;
   - View Patients' personal information;
5. Insurance Companies:
   - View Medical Records.

8.3.2 DATA MODEL

As a result, all user data and medical records will be kept in the blockchain. So, in order to distinguish between various business assets, four chaincodes were designed for this purpose:

1. User information;
2. Patient medical record;
3. Healthcare entity medical record;
4. Emergency medical record.

Table 8.5 depicts the first asset, User information, which represents all of a user's personal data. Those final two variables in Table 8.5 are only necessary for the patient's initial information and previous medical conditions, respectively. Table 8.6 shows the patient's personal data. A Patient's medical history is stored in the background variable.

Forms[55–57] from various insurance firms served as models for deciding on Personal and Background data. We have established a whole new asset class for representing patient medical records stored on the blockchain, which is shown in Table 8.7. One of the first steps in sharing a medical record was putting together an Access Control List (ACL), which specifies who can see a certain file or document.

TABLE 8.5 User Information.

Variable type	Variable	Description
String	Name	User's full name
String	Type	User's type (Patient, Doctor, Health Aid, Insurance Company)
Int	ID	Unique ID, like a patient number, medical unique identifier and insurance code
String	Birthdate	User's Birth Date
Into	Contact	Number for contact
Personal	Info	Patient's initial info
Background	Background	Patient's background diseases

TABLE 8.6 Personal Information

Variable type	Variable	Description
String	Gender	Patient's gender
String	Height	Patient's height
String	Weight	Patient's weight
String	Blood Pressure	Patient's blood pressure
String	Alcohol	Alcohol consumption
String	Smoke	If the patient is a smoker
String	Blood Type	Patient's blood type

TABLE 8.7 Patient Medical Record Information.

Variable type	Variable	Description
Int	Record ID	Unique random generated number
String	Medical ID	ID of the medical staff that is inserting that record
String	Patient ID	Patient's ID
String	Date	Record's insertion date and time
String	Specialty	Medical specialty
String	Local	Where the patient was treated
String	Notes	Medical annotations
String	Document	Extra medical documents encoded in base 64
String	Document Name	Extra medical document's name
AccessControl	Access Control	Access control list with the users' ID which this record was shared with

Medical records for Healthcare Entities differ slightly from those of patients in that they do not contain ACL information. The ACL does not present in the clinical file of the healthcare institution because this type of information can only be retrieved by doctors, health aids, insurance firms, or emergency services.

The emergency documents have the same dataset as the medical file and the patient's original details. For emergency services, having an ACL to track the medical personnel with whom a patient's data was communicated is essential.

8.3.3 USE CASES

Four different use cases were developed to demonstrate the characteristics of the proposal in order to verify the system's architecture:

a. Registration and Authentication of user
b. Electronic Health Records (EHRs) insertion and display.
c. Make use of electronic health records (EHRs).
d. The state of emergency.

The User Registration and Authentication use case is defined in Figure 8.7. During the user registration step, the user enters all of the information outlined in Table 8.5, and MedicalBlockchain registers the user in the HLF network and creates a block containing the user's initial information. At the conclusion of this operation, the HLF network's issued password will be obtained. However, during this procedure, cryptographic material is generated in order to enable the user to initiate transactions. The user will enter his or her unique ID and password during the authentication step, which will be validated by MedicalBlockchain. MedicalBlockchain's verification process verifies that the user is enrolled in the software and that his cryptographic information is genuine.

Figure 8.8 depicts a use case for inserting and viewing medical records. In this scenario, the patient will provide the doctor with his personal unique ID number. Internally, two different sorts of data structures are formed when the Doctor inserts the information shown in Table 8.7. It is expected that the patient's medical records will be the first transaction type. The second type is a Medical Record for a Healthcare Entity that can only be accessed by the physician who generated it. Patient and Doctor keys are used to encrypt the medical records of patients and healthcare organizations. In order to examine his medical records, the Patient will be notified after the transaction is completed.

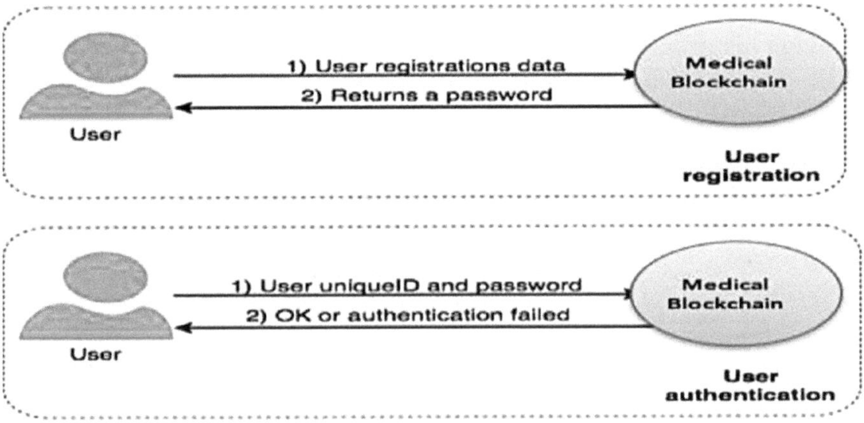

FIGURE 8.7 Registration and authentication of user.

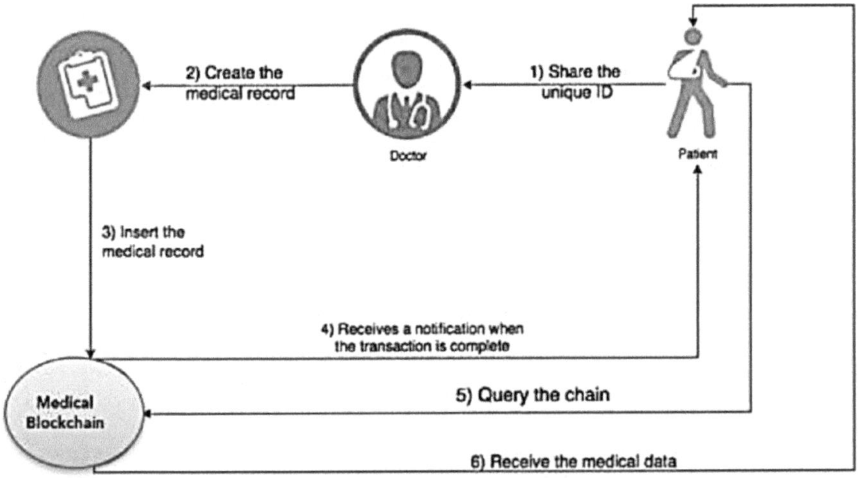

FIGURE 8.8 Electronic health records (EHRs) insertion and display.

Medical records can be easily shared using the use case shown in Figure 8.9. Here, the patient asks for Doctor 2 to have access to his medical records. An internal query of the patient's medical record adds Doctor 2's ID to the access control list, and a brand-new transaction based on the healthcare organization medical record data structure is produced with the key of Doctor 2. The doctor can inspect the transaction once it has been completed.

Finally, as seen in Figure 8.10, the emergency state functionality is demonstrated. A user's consent to having his or her personal information shared with emergency services is indicated during registration in the app. Having his or her medical history available to a doctor will help expedite therapy in this situation.

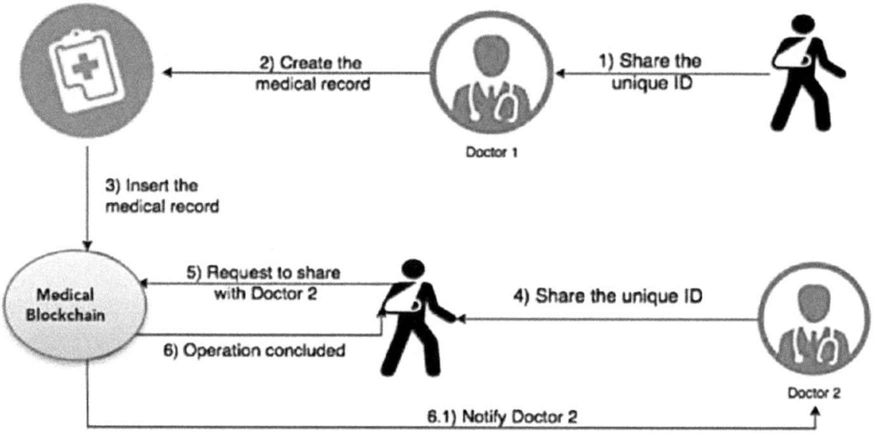

FIGURE 8.9 Make use of electronic health records (EHRs).

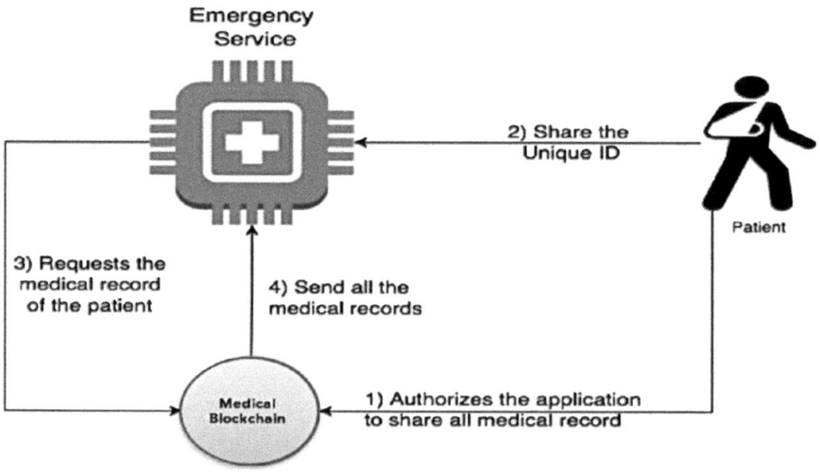

FIGURE 8.10 The state of emergency.

8.4 SYSTEM IMPLEMENTATION

This section focuses on MedicalBlockchain's architecture. A more thorough explanation of each component and its operation will be provided in the next sections.

Figure 8.11 shows the four primary technological components of MedicalBlockchain.

Technically, the MedicalBlockchain Hyperledger Fabric network is the back-ledger office's service provider.[58] YAML is a data serialization format for defining network components. To replicate the network, we use Docker containers to create a virtual version. Every business requirement is specified in the Go Programming language,[59] which is used to store the blockchain data structure.

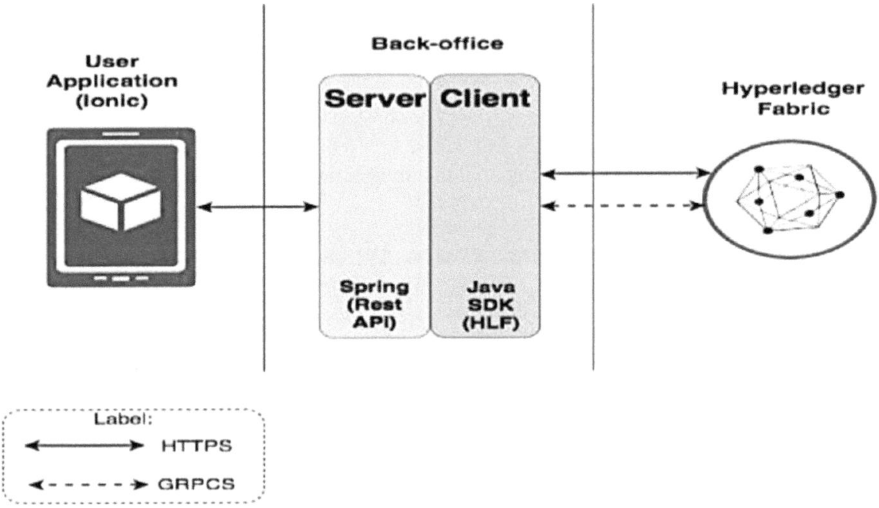

FIGURE 8.11 MedicalBlockchain's components.

Back-office takes queries from the user application and transmits them to the Hyperledger Fabric network. As seen in Figure 8.12, this solution is split into two parts. Implementation of REST API for the user application was done using the Spring Boot framework. The SwaggerUI platform was also used to construct a user interface. The Java SDK for Hyperledger Fabric version 1.1[62] was used to develop the Back-second office's component. In order to communicate with Hyperledger Fabric's network, the Back-office was developed.

Last but not the least, MedicalBlockchain includes a user application as a demonstration of the proposed scheme. The user application contains all the features necessary to serve the defined use cases. In order to create cross-platform applications, the user application was built utilizing the Ionic framework.[63] Simulate certain use cases across a variety of platforms

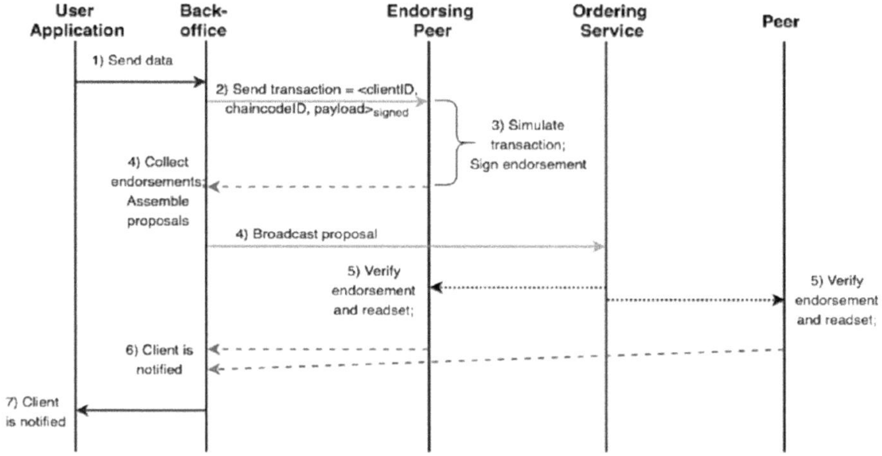

FIGURE 8.12 Multiple parties engaged in a transaction in MedicalBlockchain.

in this manner. The Ionic3 framework, Typescript, and HTML were used to construct this application.

8.4.1 MEDICALBLOCKCHAIN TRANSACTION FLOW

i. The user enters the information requested by the mobile app.
ii. The Back-office receives this data via a REST request.
iii. The transaction will include the user ID, the Chaincode ID, and the information obtained from the mobile application in the Back-office transaction. The Hyperledger Fabric CA created a User certificate when the User enrolled in the application and will be used to digitally sign the transaction. In the next step, the transactions will be sent to the people who endorsed them; thereafter.
iv. An endorsement signature is generated by the endorsing peers in a reenactment of the transaction. They examine the chaincode's access control restrictions to see if the client is authorized to make the transactions. As a result, the current situation of affairs is taken into account. As a consequence of this operation, endorsing peers will send the outcome to the client (the read and write sets connected with this present state) together with the signature of the endorsing peer. The ledger is not modified at this moment.
v. An endorsement transaction is assembled by the Back-office, which verifies that the endorsing peers have signed their names

and if the answers have the correct read/write permissions, as well as fulfilling the endorsement criteria. An envelope containing the signatures and Channel ID of the endorsing peers, as well as a read/write set, will be produced if all criteria are met.

vi. The service api receives a broadcast of the transaction proposal when it has been prepared. Envelopes are solely gathered from all channels in the network, ordered via atomic broadcast, and subsequently signed chain blocks holding these envelopes are created by the ordering service.

vii. After a block is constructed, it is delivered to all of the peers, including those who have endorsed the block. After receiving the block, peers check to see if the endorsing criteria were met and if the read set has not been altered after the transaction was completed. Transaction proposals contained in envelopes are identified as legitimate or invalid based on the results of these checks.

viii. Upon receiving a block, peers will add it to the blockchain. In order to let the Back-office know that the transaction has been added to the channel's blockchain permanently, an event is generated.

ix. User application finally gets a notification about the new transactions.

8.5 RESULTS

MedicalBlockchain's performance, fault tolerance, and functionality have been evaluated in this area. To put MedicalBlockchain through its paces, a testing environment is described in Section 3.5.1. According to Section 3.5.2, various configurations and supported operations are tested in terms of performance and latency. In this section, we look at how the adopted ordering service handles tolerated failures. As part of Section 3.5.4, the supported system's functionalities and user functionalities are described, as well as the performance benchmarks.

8.5.1 TESTING ENVIRONMENT AND EVALUATION CRITERIA

For evaluating MedicalBlockchain, a Workstation is running Debian 9 with a discrete Intel Xeon D-1700 processor and 128GB of DDR4 RAM. With the help of this specialized server (HLF Server), we were able to test several configurations of the Hyperledger Fabric network.

For testing MedicalBlockchain's functionality, we employed the Ubuntu Server (Back-office Server) with a 2.50 GHz Intel Core i7 and 12 GB DDR3 of RAM.

Three sorts of tests were performed to assess this proposal:

1. Optimal network configuration is the primary purpose of ***Performance testing***. In these testing, the primary goal was to find the best combination of block size in bytes and the number of transactions per block to optimize throughput and reduce latency. Hyperledger Fabric blocks are formed in a variety of ways, and this test demonstrates that. Afterward, the effect of varying the number of peers and orderers was studied further. Data ciphering costs were also investigated in this part. A write, a read, an update, and a delete operation were used to calculate the throughput and latency numbers. Performance comparisons are also made between running the Hyperledger Fabric network in the HLF Server and the Back Office Server.
2. Testing the system's ability to handle errors is based on the impact on throughput and latency when increasing the number of Kafka brokers ***Fault tolerance*** test is performed.
3. ***Functional testing*** focuses on ensuring that all of a component's features are working properly.

The typical network architecture is 2 peers, one of which is the endorsing peer, one ordering service, four Kafka brokers, and three zookeeper servers in all of the completed tests.

Hyperledger Caliper, a blockchain benchmarking tool, was used to measure metrics including Transactions per Second (TPS) and Latency (latency) for the Hyperledger Fabric Network benchmark and performance tests. It is important to note that because this platform is currently under development, it is unable to enable a network configuration with more than one ordering service node.

8.5.1.1 PERFORMANCE TEST

a. *Hyperledger Fabric Network benchmark*

HLF network speed and latency might be affected by the block size, according to Ref. [75]. First, the throughput and latency were measured using the default approach, which equates to 512 KB of maximum block size and 10 numbers of transactions per block. Figure 8.13 illustrates the multiple block sizes that were examined for the following findings and the evaluation of the block size's significance in MedicalBlockchain. 0.5, 1, 2, 4, 6, and 8MB blocks were all examined.

There is no improvement in throughput or latencies after 6MB as seen in both graphs (Figure 8.13). In order to get the highest possible throughput, the subsequent experiments used a 6MB block size.

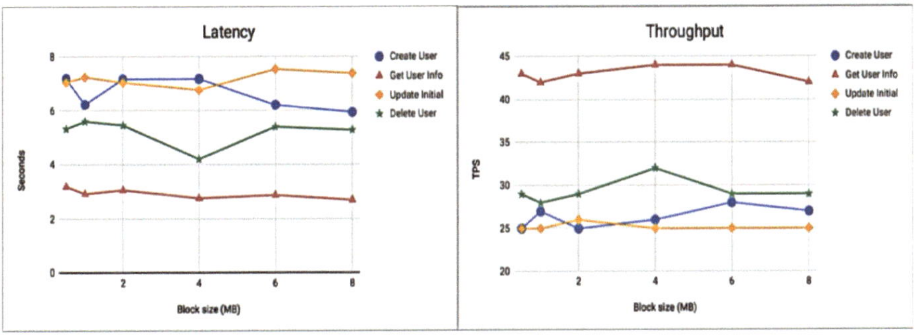

FIGURE 8.13 Different block sizes impact.

After allocating a block size, the influence of message volume on performance and delay was examined. There are 50 transactions per block for a desired block size of 6MB in Figure 8.14, which reduces delay and maximizes throughput.

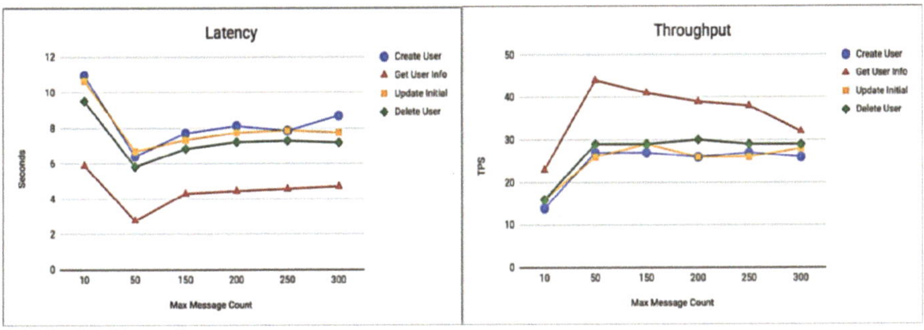

FIGURE 8.14 Number of messages impact on the block.

A single transaction was sent to define to determine the acceptable latency for subsequent experiments (Create User (write operation): 2,78s, Get User (read operation): 2,69s, Update Initial Information of a User (update operation): 2,73s, Delete User (delete operation): 2,79s), and the results were compared to the ideal block size and number of transactions per block found during the research.

According to Ref. [75], experiments were carried out to determine the optimal block size in terms of performance and latency. Using a 2MB block size and 0.5s latency, the greatest results were achieved in these testing. These results are clearly different from the ones that were previously obtained. To explain this discrepancy, consider that the network entities in the cited study ran in separate Virtual Machines, and the chaincode utilized performs simpler operations, resulting in faster execution.

b. ***Peers and ordering service nodes impact***

To see how an increase in the HLF network's throughput and latency might affect performance after determining the ideal block-size and message-per-block values, experiments were conducted. Peer count has a significant impact on throughput and latency in Figures 8.15 and 3.16. To emulate communication latency, the network was set up on the HLF Server, and Hyperledger Caliper was run from the Back-office Server.

Because the HLF Server is a one-of-a-kind computing environment, the throughput is diminishing. Using the real-time resource utilization monitoring tool (htop), it was found that the CPU is always around 100% when the quantity of instances is greater than 15. Degradation is a significant concern when using HLF Server, as all the components of the Hyperledger Fabric network run in the identical computing instance regardless of the resources given. The peer-to-peer and node-to-node communication will be faster because all components are executing on a single system.
Because of Hyperledger Caliper's restrictions, a Java application was written to estimate network throughput and latency while adding more orderers using the create user capability. Throughput was found to be maximized with a n orderers (n) of $n = p + 1$ after experiments with a variety of peers (p) and orders (n). Throughput and delay are shown in Figures 8.17 and 8.18 when the number of orderers is varied in contrast to the equal amount of peers with

only one orderer. An order fulfilment service is not a bottleneck in this network test, as previously thought. All other Hyperledger Caliper tests will not be affected by the use of only one orderer.

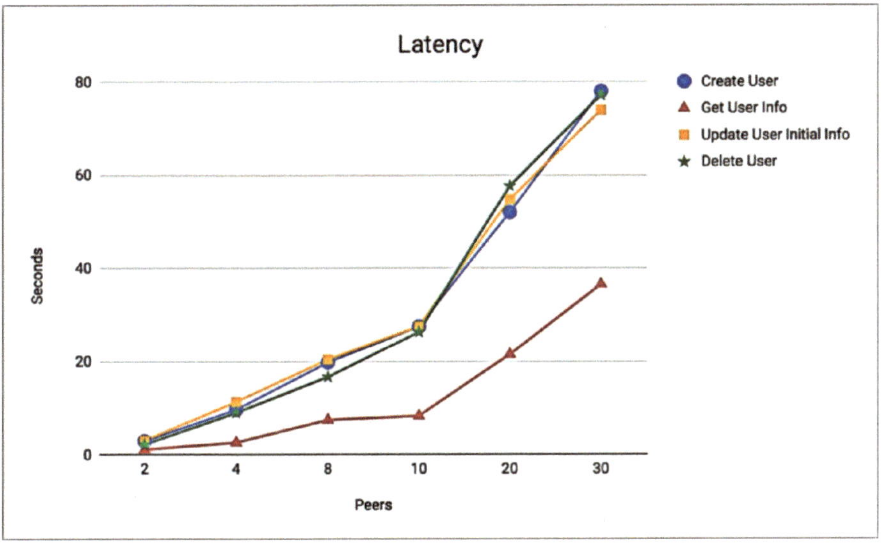

FIGURE 8.15 Impact on latency peers variation.

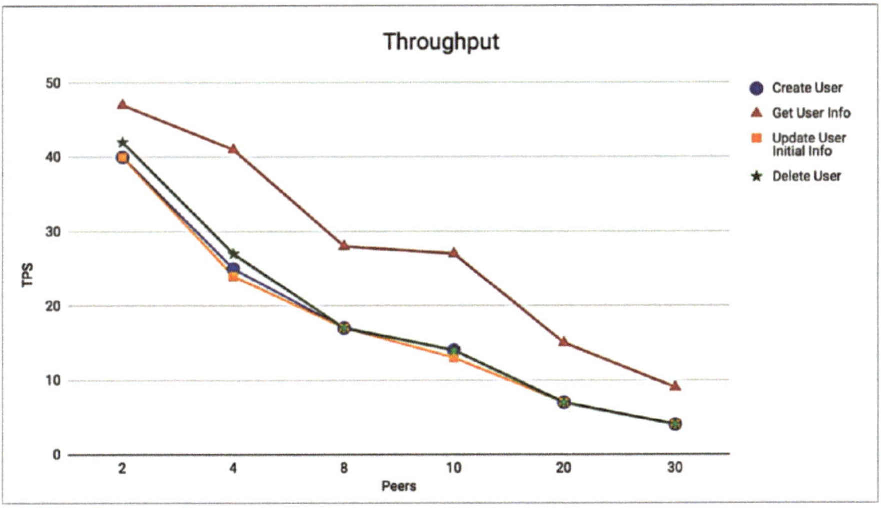

FIGURE 8.16 Impact on throughput peers variation.

Ref. [73] examined Hyperledger Fabric V1.0's performance and identified potential bottlenecks in order to gain a better knowledge of the technology and encourage future improvements. Multiple calls to CouchDB's REST API and the cryptographic operations were found to be the main performance bottlenecks by the authors.[73] The authors from the same study also observed a significant influence when the endorsement process implements various policies, varied numbers of channels, and different block sizes.

c. ***Encryption Test***

As expected, throughput is greater despite ciphering the data in this phase of the experiment. Throughput increases by 11 percent when storing data without ciphering, as shown in Figure 8.19 by the Add User function and Retrieve User Data action. While in the operation of Remove User, throughput is 10% higher, for Updating User the throughput is 13% greater than before.

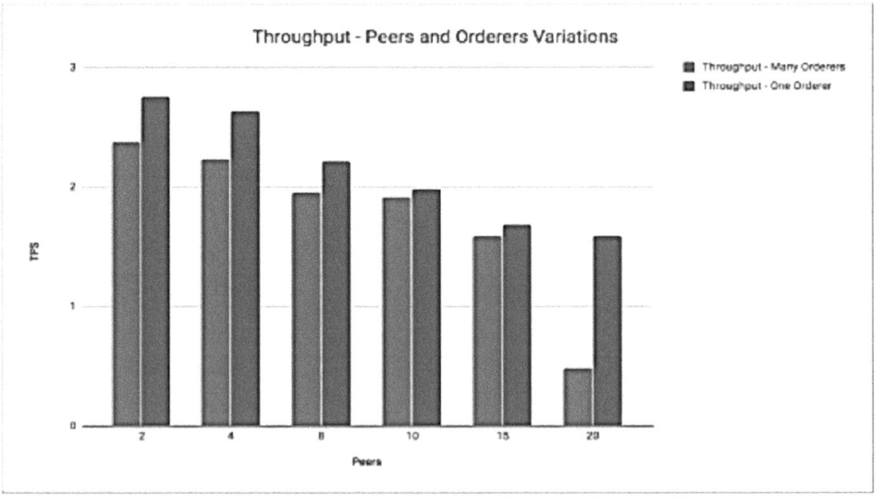

FIGURE 8.17 Throughput Peers and orderers variation.

d. **Fault Tolerance Test**

In this part, the influence on performance when the system is vulnerable to faults at the start of the execution is illustrated. The amount of Kafka brokers was changed to enable a growing amount of broker faults for these experiments. The four, seven, ten, and thirteen Kafka brokers were employed, with one, two, three, and four fault tolerances, respectively. Failure-free environments are more susceptible to delays caused by using a variable number of brokers, as shown by Figure 8.21. Consensus processing functions are more efficient because there are fewer follower brokers to keep in sync with the leader broker, resulting in fewer internal messages.

e. **Functional Tests**

Nine use-cases were developed to test MedicalBlockchain's functionality. To run these tests, we used JUnit. First, there were four doctors and one patient in the system. The following is a breakdown of the fines for each usage case.

Use-Case1—Create medical record:

- Doctor 1 generates a patient medical record.
- The medical record can be viewed by both the doctor and the patient.

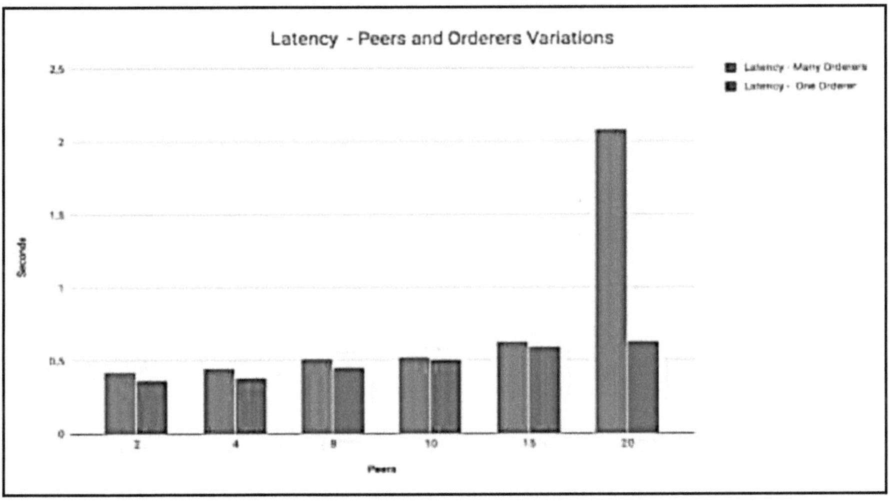

FIGURE 8.18 Latency peers and orderers variation.

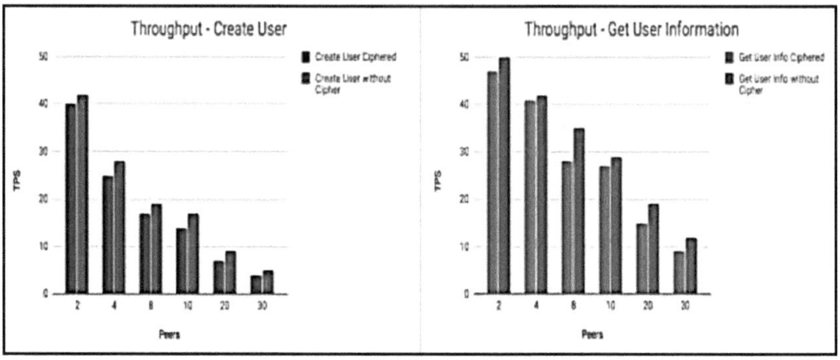

FIGURE 8.19 Create and get user.

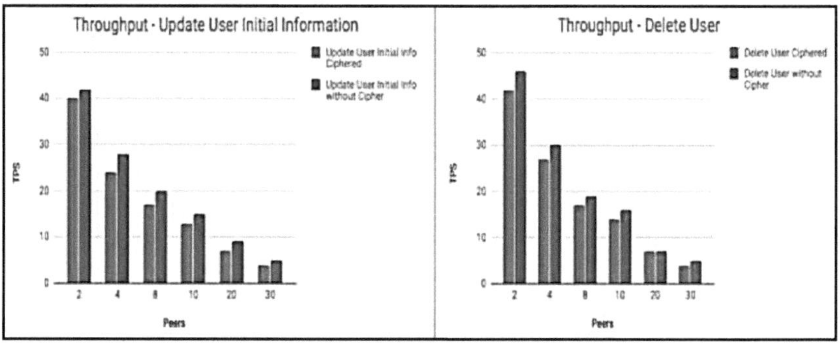

FIGURE 8.20 Update user initial information and delete user.

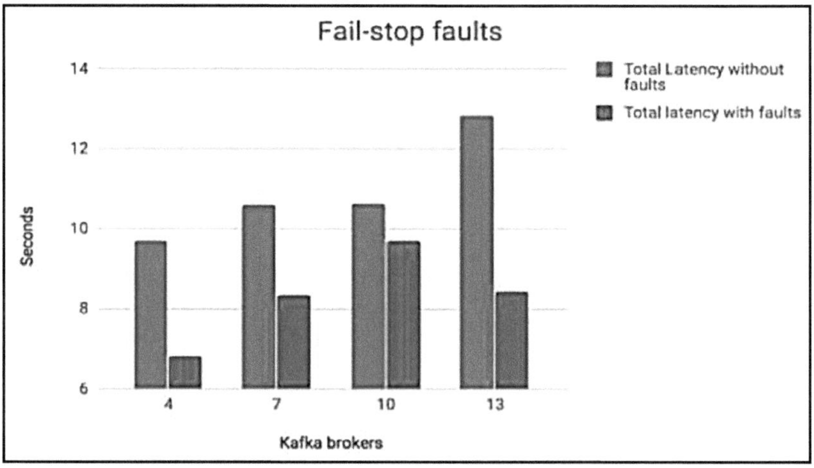

FIGURE 8.21 Fail stop comparison.

Use-Case2—Share medical records:

- Doctor 1 generates a patient medical record.
- Doctor 1 creates a medical record for the patient, which the patient exchanges with Doctor 2.
- Doctor 2 is able to see the patient's medical file.

Use-Case3—Share Initial information with emergency service:

- Patient reveals the primary information with the emergency service.
- Patient's preliminary information can be seen by Doctor 3 (an emergency service doctor).

Use-Case4—Share medical records with Emergency service:

- Doctor 2 generates a patient medical record.
- Patient exchanges the medical records with the emergency service.
- Doctor 3 (working in the emergency service) is able to see the patients' medical file.

Use-Case5—Patient deletes medical record:

- Patient generates a medical record.
- Patient erases the medical record generated by him.

Use-Case6—Patient tries to delete a medical record:

- Doctor 2 generates a patient medical record.
- There was no success in deleting EMR for the patient because it was an unauthorized action to perform.

Use-Case7—Withdraw access privileges:

- Doctor 2 generates a patient medical record.
- Doctor 2 creates a medical record for the patient, which the patient exchanges with Doctor 5.
- Doctor 5 is able to see the patient's medical file.
- Patient removes the Doctor 5 ID from their list of authorized users.
- Doctor 5 is not able to see the patient's medical file.

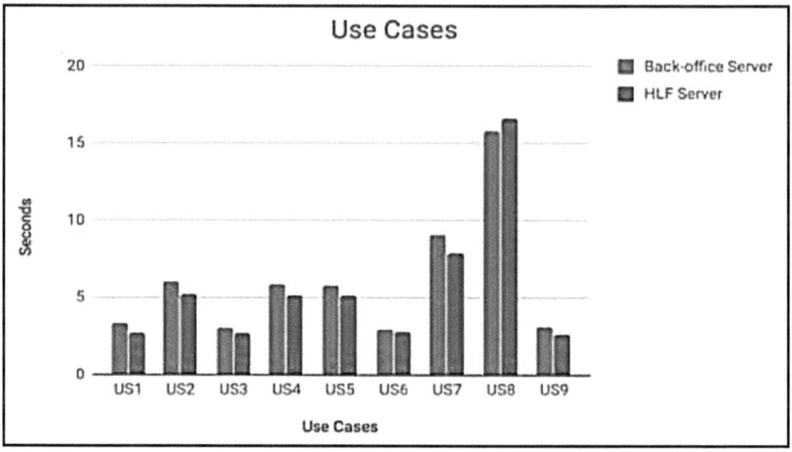

FIGURE 8.22 Back-office server vs HLF server.

Use-Case8—Delete Patient Account:

- Patient 2 should be added to the list.
- Doctor 2 generates a patient 2 medical record.
- Delete the account of patient 2.
- Doctor 2 is not able to see the medical file.

Use-Case9—Create emergency medical record:

- Doctor 4 records the patient's medical history in an emergency medical file.
- Patient, Doctors 3 and 4 is able to see the emergency medical file.

JUnit tests were also conducted on a computer with the network operating on the HLF server to evaluate the latency values reported when a user completes a task in Back Office Server. Figure 8.22 shows that when the HLF network is operating in a dedicated server, the overall time to perform each use case is reduced. That is how we're able to deduce 8% less latency can be achieved when executing the HLF network in the HLF Server instead of a single workstation.

8.6 CONCLUSION

This part summarizes the chapters' major results, as well as any remaining unresolved concerns. It also offers some potential directions for further research.

One of the most valuable things today is information, which is why it is so dispersed throughout so many different organizations. Concerns about service uptime and storage security abound with this much data, and these issues have the potential to compromise user privacy.

It has become increasingly difficult to keep track of a patient's medical history in a single location. The fragmentation of health records is a major issue in healthcare services.[3] There may be discrepancies in diagnoses between different doctors because patients cannot accurately describe what one doctor stated and because there are no procedures for exchanging information between medical institutions and institutions themselves. An electronic health record (EHR) system that allows all participants to access and edit EHRs is critical to ensuring an accurate diagnosis and effective treatment.

Ultimately, the main purpose of this chapter was to make it possible for patients to access their electronic health records, which had been created by a variety of healthcare providers, at any time. To store information that can be read, added or modified, shared or deleted by specified entities based on defined access policies is permitted.

It is possible to maintain all of the patient data with blockchain technology and ensure that it cannot be altered or deleted, but it can be modified from a user's viewpoint by uploading new data to the chain. This technology helps to eliminate data fragmentation. It is possible to share health records across organizations using the blockchain without sacrificing data security or the privacy of individual owners.

Due to defined access control constraints, a consortium blockchain was found to be most effective, since it allows the definition of access permission kinds. An in-depth look at numerous implementations of blockchain technology was the goal of the primary blockchain. When determining which platform would be best for the proposed application's development, various decentralized application development platforms were thoroughly examined. Health applications were examined in order to gather the most beneficial features from many sources for future study and consideration. Hyperledger Fabric was found to be the best suitable platform for the implementation of the suggested solution following an evaluation of the development environments and health apps available.

Detailed information about MedicalBlockchain may be found in the System Model and Implementation sections. The functionality, data schema, and supplied use cases are all extremely extensive. Three layers of the MedicalBlockchain's application architecture were used to identify the HLF network, the back-office, and user application. The proposed architecture and implementation shows the key aims of this chapter, which were to give a comprehensive solution to the challenges of data fragmentation in Health applications while ensuring data and user privacy. Communication and storage of data across system components were given extra consideration in this concept due to privacy concerns.

It was decided to run three different kinds of trials to see how well the proposed framework would perform. To find the best network configuration, a series of tests were conducted. For Fault Tolerance, increasing the number of Kafka brokers and the amount of failures that can be tolerated by the system might have an impact on throughput and latency. Finally, the primary purpose of functional tests is to verify each and every feature. When compared to other potential alternatives, the proposed solution's performance results were disappointing. Using Blockchain technology to solve the healthcare data fragmentation problem is a possible alternative, according to these findings. The immutability and logging features of the blockchain give consumers with a platform for knowing what has been done with their data and alleviate trust issues associated with decentralized environments.

8.7 FUTURE DIRECTIONS

Only fail-stop tolerance is provided by the Kafka service that was selected for implementation. A Byzantine Fault Tolerant method like BFT-SMaRt[31] could be used to investigate the effectiveness of Hyperledger Fabric's modularity.

AES was the sole encryption technique employed due to the constraints of the encCC library used to secure sensitive information. It would be fascinating to experiment with different encryption techniques to see how they affect the performance of the system. Keys supplied via a transient property and a key repository should be used for encryption, but this has not been addressed.

Patients should be able to disclose all of their clinical records at once in an emergency circumstance, but this has yet to be addressed.

Although the proposed method yielded fascinating results, there are still ways to improve the system. Delloite's[85] study shows that implementations should have both on-chain and off-chain storage. Off-chain storage should be used for large data, such as X-rays, because on-chain storage's limited scalability will be compromised by the data's large block size. In order to evaluate the impact of an off-chain storage system to one that relies solely on on-chain storage, it would be fascinating to develop and evaluate this option.

Another issue that went unaddressed was the use of authentication methods. A token-based authentication scheme like OAuth or OpenID could be used, or *Cartão de Cidadão* API for example. The personal attributes of doctors and other entities should be used to verify their legitimacy.

KEYWORDS

- **blockchain**
- **hyperledger fabric**
- **healthcare**
- **decentralized**
- **electronic health record (EHR)**

REFERENCES

1. *Processo Clinico*. http://portalcodgdh.min-saude.pt/index.php/Processo_clinico.
2. Electronic Health Record. http://portalcodgdh.min-saude.pt/index.php/Electronic_Health_Record(EHR).
3. Medicalchain Whitepaper. https://medicalchain.com/en/whitepaper/, 2017.
4. Sns. https://servicos.min-saude.pt/utente/.
5. *Hospital da Luz*. https://portalcliente.luzsaude.pt/pt.
6. *Lus'ıadas*. https://omeuportal.lusiadas.pt/pt/Paginas/Login.aspx.
7. Hearn, M. Corda: A Distributed Ledger, 2016.
8. Nakamoto, S. *Bitcoin: A Peer-to-Peer Electronic Cash System*; 2009.
9. Swan, M. *Blockchain - Blueprint for a New Economy*, 1st ed.; O'Reilly Media, Inc., 2015.
10. Antonopoulos, A. M. *Mastering Bitcoin: Unlocking Digital Crypto-Currencies*, 1st ed.; O'Reilly Media, Inc.; 2014.
11. Lin, I.-C.; Liao, T.-C. A Survey of Blockchain Security Issues and Challenges. *I. J. Network Secur.* **2017,** *19,* 653–659.
12. Yip, K. *Blockchain and Alternative Payment Models*; 2017.

13. Lamport, L.; Shostak, R. E.; Pease, M. C. The Byzantine Generals Problem. *ACM Trans. Program. Lang. Syst.* **1982,** *4,* 382–401.
14. Pilkington, M. *Blockchain Technology: Principles and Applications.*
15. On Public and Private Blockchains. https://blog.ethereum.org/2015/08/07/on-public-and-private-blockchains/.
16. Sankar, L. S.; Sindhu, M. R.; Sethumadhavan, M. In *Survey of Consensus Protocols on Blockchain Applications*, 2017 4th International Conference on Advanced Computing and Communication Systems (ICACCS), 2017; pp 1–5.
17. King, S.; Nadal, S. *Ppcoin: Peer-to-Peer Crypto-Currency with Proof-of-Stake*; 2012.
18. Vukolic, M. The Quest for Scalable Blockchain Fabric: Proof-of-Work vs. bft Replication, in *iNetSeC*, 2015.
19. Zheng, Z.; Xie, S.; Dai, H.; Chen, X.; Wang, H. In *An Overview of Blockchain Technology: Architecture, Consensus, and Future Trends*, 2017 IEEE International Congress on Big Data (BigData Congress), 2017; pp 557–564.
20. Massimo, D. P. What is the Blockchain? *Comput. Sci. Eng.* **2017,** *19* (5), 92–95.
21. Herlihy, M. *Blockchains from a Distributed Computing Perspective*; 2018.
22. Back, A. *Hashcash—A Denial of Service Counter-Measure*; 2002.
23. Buterin, V. *Ethereum White Paper*; 2014.
24. John, R.; Cherian, J. P.; Kizhakkethottam, J. J. In *A Survey of Techniques to Prevent Sybil Attacks*, 2015 International Conference on Soft-Computing and Networks Security (ICSNS), 2015; pp 1–6.
25. Christidis, K.; Devetsikiotis, M. Blockchains and Smart Contracts for the Internet of Things. *IEEE Access* **2016,** *4,* 2292–2303.
26. Saleh, F. *Blockchain Without Waste: Proof-of-stake*; 2018.
27. Nxt Whitepaper. https://web.archive.org/web/20150203012031/http://wiki.nxtcrypto.org/wiki/Whitepaper:Nxt#Blocks.
28. Vasin, P. Blackcoin's Proof-of-Stake Protocol v2, 2014.
29. Casper. https://github.com/ethereum/wiki/wiki/Casper-Proof-of-Stake-compendium.
30. Castro, M.; Liskov, B. Practical Byzantine Fault Tolerance and Proactive Recovery. *ACM Trans. Comput. Syst.* **2002,** *20,* 398–461.
31. Sousa, J.; Bessani, A. N.; Vukolic, M. In *A Byzantine Fault-Tolerant Ordering Service for the Hyperledger Fabric Blockchain Platform*; 2018 48th Annual IEEE/IFIP International Conference on Dependable Systems and Networks (DSN), 2017; pp 51–58.
32. Schwartz, D.; Youngs, N.; Britto, A. *The Ripple Protocol Consensus Algorithm*; 2014.
33. Kwon, J. *Tendermint: Consensus Without Mining*; 2014.
34. Conti, M.; Kumar, S.; Lal, C.; Ruj, S. A Survey on Security and Privacy Issues of Bitcoin. *CoRR* 2017, vol. abs/1706.00916.
35. Dinh, T. T. A.; Wang, J.; Chen, G.; Liu, R.; Ooi, B. C.; Tan, K.-L. In *Blockbench: A framework for Analyzing Private Blockchains*, *SIGMOD Conference*, 2017.
36. Solidity Documentation. http://solidity.readthedocs.io/en/develop/.
37. Luu, L.; Chu, D.-H.; Olickel, H.; Saxena, P.; Hobor, A. Making Smart Contracts Smarter. *IACR Cryptology ePrint Archive* **2016,** *2016,* 633.
38. Quorum Whitepaper. https://github.com/jpmorganchase/quorum-docs/blob/ master/Quorum%20Whitepaper%20v0.1.pdf, 2016.

39. Chain Protocol Whitepaper. https://chain.com/docs/1.2/protocol/papers/ whitepaper, 2018.
40. Brown, R. G.; Carlyle, J. R.; Grigg, I.; Hearn, M. Corda: An Introduction, 2016.
41. Corda. https://docs.corda.net/index.html.
42. Hyperledger Burrow. https://github.com/hyperledger/burrow.
43. Iroha Whitepaper. https://github.com/hyperledger/iroha/blob/master\/docs/ iroha_whitepaper.md.
44. Hyperledger Swatooth. https://sawtooth.hyperledger.org/docs/core/releases/ latest/introduction.html.
45. Hyperledger Fabric. https://hyperledger-fabric.readthedocs.io/en/latest/ blockchain.html.
46. Monax. https://monax.io.
47. Proof of Elapse Time. https://sawtooth.hyperledger.org/docs/core/releases/latest/architecture/poet.html.
48. Kafka and Zookeeper. https://hyperledger-fabric.readthedocs.io/en/release-1.1/kafka.html#kafka-caveat.
49. Glossary. https://hyperledger-fabric.readthedocs.io/en/release-1.1/glossary.html.
50. Ekblaw, A. C. MedRec: Blockchain for Medical Data Access, Permission Management and Trend Analysis [PDF MedRec: Blockchain for Medical Data Access, Permission Management and Trend Analysis], 2017. https://www.media.mit.edu/publications/medrec-blockchain-for-medical-data-access-permission-management-and-trend-analysis/
51. Dubovitskaya, A.; Xu, Z.; Ryu, S.; Schumacher, M.; Wang, F. In *Secure and Trustable Elec-Tronic Medical Records Sharing Using Blockchain*, AMIA ... Annu. Symp. Proc. AMIA Symp. **2017,** vol 2017, pp 650–659.
52. Ahram, T. Z.; Sargolzaei, A.; Sargolzaei, S.; Daniels, J.; Amaba, B. In *Blockchain Technol- ogy Innovations*, 2017 IEEE Technology & Engineering Management Conference (TEM- SCON), 2017; pp 137–141.
53. Ibm Cloud. https://www.ibm.com/cloud-computing/bluemix/.
54. General Data Protection Regulation (GDPR). https://gdpr-info.eu.
55. Millennium. https://ind.millenniumbcp.pt/pt/Particulares/insurance/Documents/Questionario_MedisFIT.pdf.
56. Fidelidade. https://www.fidelidade.pt/PT/particulares/Familia/Seguros/Documents/SeguroVida-Proposta+IPC.pdf.
57. Real Vida Seguros. https://www.realvidaseguros.pt/images/stories/pdf/real_vida_previdencia_questionario_clinico.pdf.
58. Yaml. http://yaml.org.
59. Golang. https://golang.org.
60. Spring. https://spring.io.
61. Swagger. https://swagger.io/tools/swagger-ui/.
62. Java Sdk for Hyperledger Fabric. https://github.com/hyperledger/fabric-sdk-java.
63. Ionic Framework. https://ionicframework.com.
64. Fabric ca Glossary. https://hyperledger-fabric.readthedocs.io/en/release-1.1/glossary.html#hyperledger-fabric-ca.
65. Public Key Infrascture. https://www.digicert.com/pki/.
66. Fabric ca. http://hyperledger-fabric-ca.readthedocs.io/en/latest/ users-guide.html.

67. Sqlite vs Mysql vs Postgresql: A Comparison of Relational Database Man agement Systems. https://www.digitalocean.com/community/tutorials/sqlite-vs-mysql-vs-postgresql-a-comparison-of-relational-database-management-systems.
68. Openssl. https://www.openssl.org.
69. Ledger. https://hyperledger-fabric.readthedocs.io/en/release-1.1/ledger. html.
70. Couchdb. https://hyperledger-fabric.readthedocs.io/en/release-1.1/couchdb_as_state_database.html.
71. Json. http://json.org.
72. Couchdb vs Leveldb. https://db-engines.com/en/system/CouchDB%3BLevelDB.
73. Thakkar, P.; Nathan, S.; Vishwanathan, B. Performance Benchmarking and Optimizing Hyperledger Fabric Blockchain Platform. *CoRR* **2018,** abs/1805.11390.
74. Sousa, J.; Bessani, A.; Vukolic, M. A Byzantine Fault-Tolerant Ordering Service for the Hyperledger Fabric Blockchain Platform.
75. Androulaki, E.; Barger, A.; Bortnikov, V.; Cachin, C.; Christidis, K.; Caro, A. D.; Enyeart, D.; Ferris, C.; Laventman, G.; Manevich, Y.; Muralidharan, S.; Murthy, C.; Nguyen, B.; Sethi, M.; Singh, G.; Smith, K.; Sorniotti, A.; Stathakopoulou, C.; Vukolic, M.; Cocco, S. W.; Yel-lick, J. Hyperledger Fabric: A Distributed Operating System for Permissioned Blockchains. In *EuroSys*, 2018.
76. Cryptogen. http://hyperledger-fabric.readthedocs.io/en/release-1.1/ commands/cryptogen-commands.html#.
77. Configtx. http://hyperledger-fabric.readthedocs.io/en/release-1.1/commands/configtxgen.html.
78. Grpc. https://grpc.io.
79. Peers. https://hyperledger-fabric.readthedocs.io/en/latest/peers/peers. html.
80. Mongodb. https://www.mongodb.com.
81. Chaincode Tutorials. https://hyperledger-fabric.readthedocs.io/en/release-1.1/chaincode.html.
82. Cid. https://github.com/hyperledger/fabric/tree/release-1.1/core/ chaincode/lib/cid.
83. Enccc. https://github.com/hyperledger/fabric/tree/master/examples/chaincode/go/enccc_example.
84. Ionic Storage. https://ionicframework.com/docs/storage/.
85. Krawiec, R.; Housman, D.; Mark White, M. F.; Quarre, F.; Barr, D.; Nesbitt, A.; Fedosova, K.; Killmeyer, J.; Israel, A.; Tsai, L. Blockchain: Opportunities for Healthcare 2016.
86. Feng, L.; Zhang, H.; Chen, Y.; Lou, L. Scalable Dynamic Multi-Agent Practical Byzantine Fault-Tolerant Consensus in Permissioned Blockchain. Appl. Sci. 2018, 8, 1919. https://doi.org/10.3390/app8101919.
87. Li Wang, Hui, Shao-I Chu, Jiun-Han Yan, Yu-Jung Huang, I-Yueh Fang, Shu Ya Pan, Wei-Cheng Lin, et al. 2020. 'Blockchain-Based Medical Record Management with Biofeedback Information'. Smart Biofeedback - Perspectives and Applications. IntechOpen. doi:10.5772/intechopen.94370.

CHAPTER 9

TOWARD SMART HEALTHCARE WITH THE INTERNET OF THINGS

ANJUM SHEIKH QURESHI

Rajiv Gandhi College of Engineering Research & Technology, Chandrapur, India

ABSTRACT

Internet of things (IoT) is being extensively used for a number of applications in real life. Some of them are smart home, agriculture, waste management, transportation, and Healthcare. IoT Healthcare is implemented with the help of sensors and medical devices. The IoT-enabled Healthcare monitoring devices can be used for gathering essential data like blood sugar, pulse rate, heartbeat, blood pressure, etc and transfer it to the doctors. The doctors in turn can provide opinions to the needy persons irrespective of their location. It is beneficial for the patients living in remote areas and also the critical patients who are unable to travel long distance to meet their doctors. There are a number of health parameters that can be monitored with the help of wearable. But when Healthcare is merged with IoT the chances of security threats will grow up. Security and privacy issues are very important as well as challenging in IoT Healthcare. If the healthcare system is hacked, the medical parameters can be altered by the hackers and wrong information will reach the doctors. Such incidents would be a great loss to the patient as it would be fatal and sometimes would lead to death. In this chapter, we will discuss about the applications of IoT in the healthcare industry.

IoT-Enabled Healthcare Systems: Applications, Benefits, and Challenges
Yatindra Kumar, Shashi Kant Verma, Archana Singh, Krishna Kumar, & Manoj Gupta (Eds.)

9.1 INTRODUCTION

Internet of things (IoT) is a promising technology, and its usage is increasing day by day. Its popularity can be known by the steep increase in the number of IoT devices in the last few years. It has been predicted that in few years the number of IoT devices would surpass the total global population. IoT has enabled controlling of anyone from anywhere at any time. With the help of IoT a user can control a number of things through the internet, be it a smart phone, a human being, or some home appliances. The usage of IoT has led to a number of inventions and innovations that intend to provide better and comfortable life to all of us. Due to its versatility, IoT is being used for a number of applications. Some of the real-life applications of IoT include smart home, smart agriculture, smart cities, smart parking system, smart Healthcare, etc.

Healthcare with the help of IoT generally termed as smart Healthcare is gaining popularity due to the number of benefits attached to it. It plays a significant role in various Healthcare applications with the help of sensors and wearable devices. The sensors record the Healthcare parameters and pass it to the doctors for analysis. The doctors in turn examine the health record of patient and provide suggestions as well as medicines to the beneficiaries. With the help of such system a person is able to contact the doctor from any place and similarly a doctor is able to track the patient's health from any part of the world. Smart Healthcare is a boon for the people living in remote places with less access to transport and to those who have to travel long distance to reach a nearby town or city to get an appointment from a doctor. It is also beneficial for the chronic and bedridden patients as their doctor will be able to monitor their health from his place. The IoT sensors can be used to take care of the physiological and the neurological conditions. A number of health parameters like blood sugar, pulse rate, blood pressure, ECG, etc., can be monitored with these sensors. It can thus be helpful in preventing serious conditions as the doctor will get notification of some drastic changes in parameters and he would therefore be able to contact his patients immediately and advise for remedies.

Though smart Healthcare has lot of advantages it has some challenges that has to be faced for better utilization of the technology. Cost, data accuracy, interoperability, and scalability are some of the issues faced by the healthcare systems. Data privacy and security risks are higher in these systems due to the data related to health of patients. Healthcare data is very important. Any changes made by the hackers in the health data would result in transmission of wrong information to the doctors who in turn may give some wrong suggestions or medicines to the patients. Such actions can be fatal and may even result in death in some cases.

In Section 2 of this chapter some applications of IoT Healthcare will be discussed. Section 3 will try to list out technologies used for Healthcare IoT, Section 4 will discuss about services of healthcare IoT while the benefits will be discussed in Section 5 while challenges in Section 6 and Section 7 will conclude the chapter.

9.2 APPLICATIONS OF IoT HEALTHCARE

The continuous research in the field of IoT has enabled the development of affordable medical devices, and wearable gadgets that can be used to gather the health information of patients, diagnose them for monitoring their health conditions, and generate alerts in case of emergencies.[1] There are a number of IoT devices that have made the health monitoring task easier for the medical practitioners.

9.2.1 GLUCOSE-LEVEL MONITORING

Diabetes is a kind of metabolic disease in which the body is not able to produce enough insulin. The regular monitoring of glucose levels helps in knowing the reasons behind fluctuation of glucose levels for a person. This record can be used by the doctors to prescribe medicines and suggest a diet chart for the patients. Careful management of diabetes is required to prevent the risk of complications like cardiac arrests, kidney disease, blindness, and strokes. An IoT-based glucose monitoring system discussed in Ref. [2] that uses a sensor-based device for presenting the values of glucose level in graphical and human readable form so that it can be easily understood by the users and doctors. This system as shown in Figure 9.1 uses a portable sensor device, backend system, and a gateway.

The portable sensor device consists of sensors, microcontroller, a wireless communication block, and energy harvesting as well as management components. The initial task of data acquisition is done by the microcontroller. It gathers the value of glucose through an implantable glucose sensor using a wireless inductive link receiver. Along with this it uses data link wires such as UART or SPI for collecting the values of environmental and

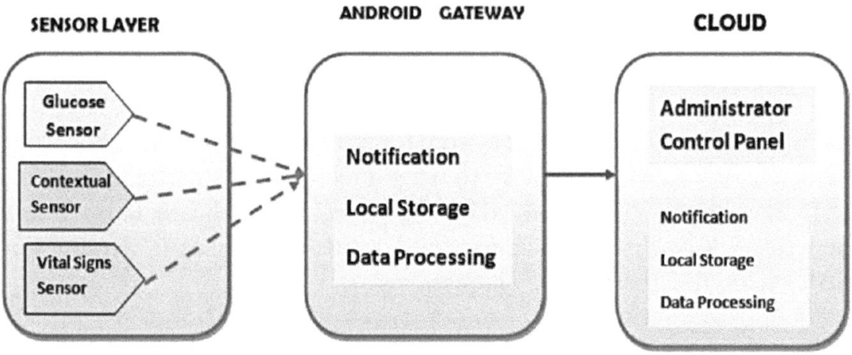

FIGURE 9.1 IoT-based glucose monitoring system.
Source: Adapted from Ref. [2]

body temperature. The data collected by the microcontroller is transmitted to the gateway with the help of a nRF wireless communication block. The gateway is equipped with a nRF transceiver. The communication block consists of a RF transceiver IC for the 2.4 GHz ISM band and an embedded antenna. The nRF can be used for both short- as well as long-range transmission and therefore the transmission range and power can be configured according to the requirement of an application. The data collected by the gateway is sent to the cloud servers. The nRF transceiver in the gateway is compatible with android phones, Iphone, and tablets. It consists of microcontroller, low-power RF Transceiver IC, and a FTDI component for the conversion of UART to USB connection. Some of the services provided by the gateway are local database, processing of data, local host with user interface, and push notifications. The data processing unit in the gateway is used for filtering noise and corrupted data, to run algorithms for processing of data to facilitate decision-making by categorization of data. The local database is comprised of intact database and real-time database. The intact database is used internally for storing algorithms and managed by the system administrators. The real-time database will store e-health and contextual data that is regularly updated and synchronized with the database of cloud. The decision-making and push notification services in the gateway work together to deliver real-time notifications to the physicians. For example, if the recorded glucose levels are detected to be lower or higher than the acceptable range the decision-making service will activate the push notifications to send messages to the doctors to inform them about the changes in real-time data. The doctor is able to access the real-time data through a web browser or with a mobile application.

9.2.2 ECG MONITORING

Electrocardiogram is a commonly used diagnostic tool for detecting heart problems and monitoring the health of heart. Regular checking of health conditions for elderly people is challenging due to the time duration needed for the checkups and also to travel to hospitals if they are located at long distances. The ECG monitoring system used by the doctors is not portable and therefore cannot be used at home for monitoring the condition of heart. Moreover in the traditional ECG monitoring systems the data has to be collected and sent to the doctor for further analysis. This process will cause delay and sudden heart failure cannot be detected. To deal with these issues, a portable ECG device has been developed by the authors in Ref. [3] that uses IoT and notifies the doctors in case of detection of medical abnormalities in the patient.

The ECG monitoring system given in Ref. [3] is divided into three parts:

a. **Sensing Network:** It consists of ECG AD8232 sensor, Aurdnio Uno, and Raspberry Pi model 2. The sensing network uses a wearable ECG sensor to collect the data physiological data from the surface of the body and transfer it to the IoT cloud using Wi-Fi. Wireless protocols like Bluetooth and Zigbee can be used but due to their limited transmission range, Wi-Fi is preferred for such systems. The AD8232 sensor collects the signals from human body using Aurdino and sends it to the Raspberry Pi. These collected signals are processed by Raspberry Pi and the final image is displayed. It uses processing software and a bandpass filter to eliminate noise in the ECG graph.

b. **IoT Cloud:** The IoT cloud is to store data and processing it. It will save the patient's information, analyze it, and generate warning messages to the patients to save them from the damages caused due to the heart diseases. An email is sent to the doctor and the uses to notify regarding the detection of abnormalities in the patient. After receiving these alerts, the doctors can take immediate actions to cure the patients.

c. **Graphical User Interface (GUI):** The GUI is used by the patients and the doctors to visualize the data. Mobile applications and web pages are the two types of GUI used in these systems for visualizing the ECG data. The users can log on to the cloud to check the collected real-time data. Mobile applications provide immediate responses, but the web pages are preferred by the users as they provide better protection of data.

An architecture of IoT-based ECG monitoring system is shown in Figure 9.2.

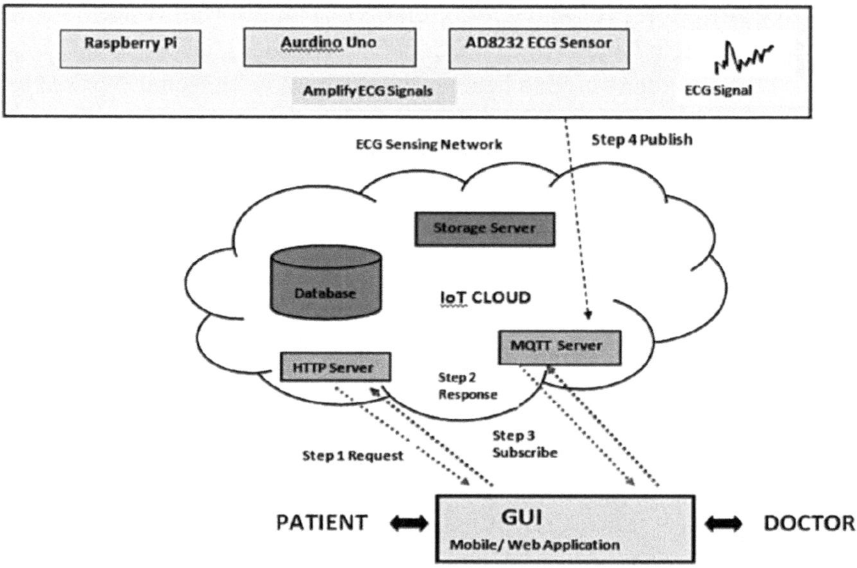

FIGURE 9.2 IoT-based ECG monitoring system.[3]
Source: Adapted from Ref. [3]

9.3 BLOOD PRESSURE MONITORING

The blood pressure should be measured at regular intervals as it keeps on fluctuating due to various reasons like stress, exertion, and extreme hot or cold weather. Though these factors cause temporary changes in the blood pressure, its monitoring is required to prevent the risks of heart attacks and strokes. Blood pressure can be measured at hospitals and at homes by using the automatic blood pressure monitoring systems. The advantage of combining IoT with these types of devices is that it will keep a continuous record of fluctuations and the person as well as the doctors will be alerted if any major changes in blood pressure are observed. One of such devices has been discussed in Ref. [4] that enables remote monitoring of blood pressure through Telegram application and emails. This device can be worn on wrist to detect the blood pressure. It is connected to a Raspberry Pi through a USB cable. The Raspberry Pi is connected to a Wi-Fi dongle for internet connection in order to transfer the data on the internet so that the users can access the data through telegram or emails. An IoT-based blood monitoring system is shown in Figure 9.3.

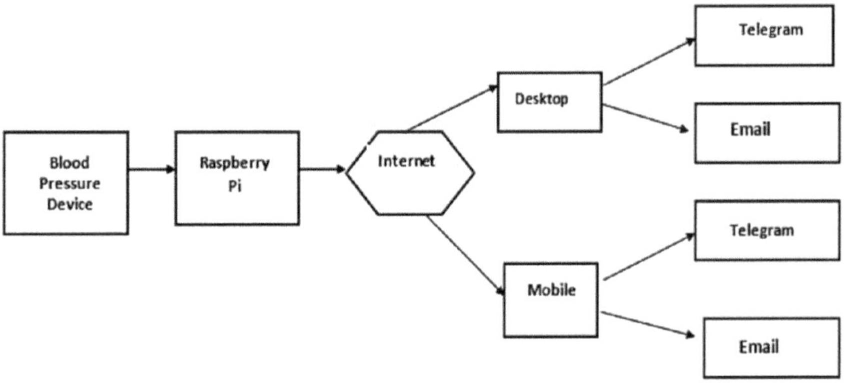

FIGURE 9.3 IoT-based blood monitoring system.
Source: Adapted from Ref. [4]

9.4 TEMPERATURE MONITORING

Measurement of body temperature is important as a number of diseases can be identified with the changes in temperature. A doctor is able to analyze the effect of medicines on a patient by keeping a track of body temperature. A fever may be caused due to infections, due to severe trauma like heart attacks and stroke or due to some other medical conditions like hyperthyroidism, arthritis, and cancer. Apart from this there is a condition called as heatstroke in which the body fails to control its own temperature and a continuous rise in body temperature is observed. It can cause dehydration and needs immediate medical attention. It is essential to check the body temperature of patients with chronic illness frequently if they are under the supervision of a doctor. Many people are unable to check their temperature accurately with the help of a thermometer. The patient or his caretakers may forget to record temperature at times and it is challenging to record temperature if the patient sleeps for a long time. In such circumstances, wearable sensors will be helpful to detect illness by continuous monitoring of body temperature.

The authors in Ref. [5] have discussed a wearable device with a Microelectromechanical system accelerometer, a digital thermometer, and a temperature sensor. This device records the temperature continuously and provides warnings if the body temperature exceeds the threshold value. This device uses contact as well as contact thermometers. The contact thermometer requires physical contact while the noncontact thermometer is an infra red sensor that is able to measure the body temperature by detecting the level of infrared emissions. This device uses a MCP9808 temperature sensor that provides benefits of low cost and stable measurement with less noise. The MCP9808 is a contact thermometer that is attached to the wearable device to measure the body temperature through physical contact. The infrared sensor is helpful in recording the body temperatures at instants when the user wants to remove the wearable for sometime or when the wearable needs to be charged due to the limited battery time. The contact thermometer gives accurate results but the infrared sensor provides the advantage of quick measurement with no contact. The recorded temperature is sent to the cloud and to the shared devices. The device uses ESP32 chip to enable wireless communication by both Bluetooth and Wi-Fi. The Bluetooth is used an option for data transfer to the devices of members at home when Wi-Fi has stopped working. The wearable uses a liquid crystal display (LCD) to display the current temperature and the alert messages when the temperature exceeds the threshold temperature.

9.5 MOOD MONITORING

The term "mood" refers to a number of feelings that impact our decision-making abilities and has a huge influence on the way we react to the situations coming our way. Mood tracking helps in managing the stress levels in an efficient manner. It is a positive psychology technique in which a person's mood is recorded at regular intervals to analyze the variations in mood of a person for certain duration of time like for a day or week. This technique helps in improving mental health. Mood tracking can help a healthy individual to stay in healthy emotional states. At the same time the technique can be helpful in health management of individuals with mental diseases like bipolar disorder and depression.

A speech emotion detection (SED) system given in Ref. [6] detects human emotions by using IoT and deep learning for care of the elderly. This system uses a SED system to record and classify the real-time audio into speech emotions. This method is helpful when the treatment of a patient depends upon their feelings in daily life. The real-time audio is recorded using a microphone which is connected to a Raspberry Pi. The recorded data is sent to the classification server. Each sound clip is trimmed into 3-s long clip after which the classification is done using librosa duration function along with data overlapping. The server classifies and predicts the emotions and gives the feedback so that the doctor can take of the patients accordingly. The classification is done using data augmentation and data normalization techniques in deep learning. The classifier uses the audio dataset to classify the recordings into different types of emotions like angry, sad, happy, disgust, fearful, surprised, and neutral.

Facial expressions can be used to study the emotional health of a person. A study in this regard has been done in Ref. [7] that uses IoT platform for individualized monitoring of elderly persons by using wireless sensor networks and smart phones. The sensors are distributed in environment that will capture images of the person and send it to the decision-maker. The decision-maker acts like an interface between the patient and helper by processing the received signals and take decisions that would help in providing treatment to the patients. The sensor elements detect the presence of a person in a room and capture images. If the detected person is not the patient, activities of that person are ignored. If the person is identified as patient all the activities are monitored. The changes in the emotional state of the individual are detected from the past and current captured images. The decision-maker uses the collected data to judge the changes in behavior of patients. If any kind of unexpected behavior is detected, the decision-maker communicates it to the patient's relative or to the caregivers.

9.6 WHEELCHAIR MANAGEMENT

The people with disabilities or older persons need wheelchair to enable their movement without help of others. The hand propelled wheelchairs are now being replaced with battery-operated or electric wheelchairs that provide more flexibility to its users. They are able to move on their own with minimal support. But to reduce the chances of miss-happenings it is essential for the family members to monitor the movement of wheelchair used by their near ones. Similarly, management of wheelchair is required by doctors and nurses when they have many persons with disabilities under their supervision. Sensors and IoT-enabled platforms can be one of the solutions for monitoring the movement of wheelchairs so that the movement of patients can be monitored continuously.

At hospitals and old age homes it is difficult for the nurses to manage many patients with wheelchair. One such IoT-enabled wheelchair management system has been given by the authors in Ref. [8] that uses sensors and Agile-Kanban. The technique known as Agile has been used to develop software while Kanban supports Agile development. This system is used to manage many wheelchairs at a time. While using electric wheelchairs it is important to know the condition of wheelchair like state of battery, and checking the parameters like voltage, current of the motor for ensuring safety of the person using a wheelchair. The combination of IoT and Kanbans is used to manage the wheelchair. The Kanban operates in different states like ready backlog, work in progress, and done. Each wheelchair requires a Kanban. When the Kanban is in backlog state it means the wheelchair requires maintenance. When the Kanban is in ready state, it means the wheelchair is ready to use. When it is in the state work in progress, it means the wheelchair is working while the state of done indicates the end of work. During working stage, the sensors record the voltage, current, and temperature on the wheelchair. All these states of wheelchair are recorded by the sensors and uploaded to the database. The doctors and the care takers can access this database to know the activity of a wheelchair eliminating the need of a person to assist the ones using wheelchair. The monitoring of wheelchairs can help in preventing the accidents that are caused due to obstacles or due to sudden stopping due to consumption of battery.

9.7 FALL DETECTION

Falls in elderly persons is very common. These falls can be due to weakness, arthritis, urinary issues, diabetes, and cardiovascular diseases and also due to environmental factors like poor lightning, improper footwear, or slippery floors. Preventing the fall of elderly persons is needed to prevent them from heavy injuries. A survey by World Health Organization (WHO) says that fall is ranked as second among the different reasons for accidental deaths while road accidents is ranked first. The accidents due to fall is found mostly among the people above 60 years of age after which they require medical attention and may even lead to long term disability.[9]

An IoT-based system will be helpful that will detect the fall and alert the family members and doctor of the elderly persons. One such approach has been presented in Ref. [10] that uses a cloud-based fall detection method for monitoring of the elderly. The researchers have developed an application called as whoops that is installed on a mobile. The mobile is used as an IoT; an IoT device that is able to sense the environment, collect, process, and transmit the data. The application uses a data centre to analyze the data for many patients and older adults. The results of the data analysis which includes the confirmation of fall of the person will be sent to the mobile in the form of alert messages. This approach uses two types of sensors with the mobile: triaxial accelerometer and triaxial gyroscope for transmitting the data of fall detection. The data sensed by these sensors is sent to the data centre in the cloud. The Azure cloud uses the IoT Hub service as the cloud gateway that helps in resolving the issues of data transmission and bidirectional connectivity between the cloud and device. The gathered data is transmitted at regular intervals and it forms a data stream which is forwarded to the stream analytics service for further processing. The data stream is processed by the stream analytics in time windows for extraction of appropriate features. When a fall is detected in the cloud it is sent to the machine learning web service to classify the situation in given period. The machine learning service determines the detection of fall and sends it back to the stream analytics module and it is served in the binary large object data store. The data is also transferred to the Azure module through HTTP request that allows defining the code of individual functions eliminating the need to create an entire server application. These defined functions perform the task of sending cloud to device messages and the result of classification through the IoT hub and the result of classification is accepted by the function from the machine learning web service. The results of the classification have two values 0 and 1. When the event is classified as fall and the result of classification is 1, an alert message is sent to the device of the caregiver in case the fall is not rejected in the application of the monitored person. In other case when the event is not classified as fall and the result of classification is 0, the event is skipped and the no message is generated or sent. A fall detection method using whoops application is shown in Figure 9.4.

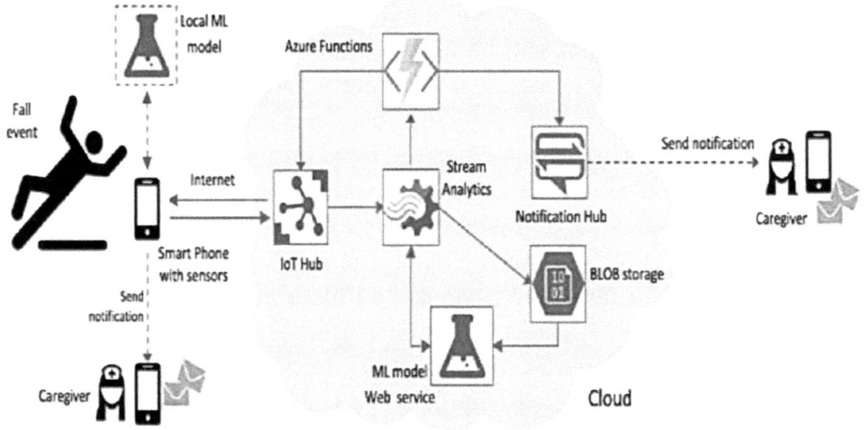

FIGURE 9.4 Fall detection method using whoops application.
Source: Adapted from Ref. [10]

9.8 MONITORING OXYGEN SATURATION LEVEL

The monitoring of the oxygen saturation levels helps on knowing the amount of oxygen in a person's blood. Ample oxygen levels in the blood indicate that the body is getting ample amount of oxygen. Generally, oximeters are used to know the oxygen levels of a person. IoT-based oximeters are helpful if the doctors have to continuously monitor the oxygen levels of chronic patients.

A pulse oximetry device that uses embedded IoT has been explained in Ref. [11] that can be used by the medical practitioner to remotely monitor the pulse and oxygen level of his patients by using a mobile phone or a computer. This IoT-based oximetry system uses a MAX30100 sensor that operates on a power supply voltage of 1.8 v and LED supply voltage of 3.3 v. The sampling rate of sensor generally used is 0.5 s which can be configured to sample faster or slower according to the external environment. It consists of an internal 16-bit sigma-delta analog to digital convertor with internal temperature compensating sensor for accurate measurement calibration. The value of oxygen levels is collected by the sensor from the patient's finger. These values are transmitted to the ThingSpeak API server through the internet router and are updated in the users account by using an API writing key and channel number ID. All the values are stored in the account channel which can be seen by the doctor at regular intervals to monitor oxygen level of a patient. The account stores the history data which can be downloaded as a CSV file and can be used for further analysis if needed. In case the doctor finds that the health of the patient is deteriorating he can inform the family members for change in medicines or advise to shift the patient immediately to the hospital.

The use of technology plays a crucial role in developing healthcare system based on IoT. The choice of software, algorithms, devices, and technology decides the success of any application. While developing an application based on IoT, the developers have to be careful in selection of three technologies: identification, communication, and location.[1]

9.8.1 IDENTIFICATION TECHNOLOGY

One of the practical considerations while working with the Healthcare IoT is accessing of a patient's data coming from an IoT node at remote locations. Effective identification of nodes is essential to make correct use of the data arriving from the nodes of healthcare network. An identifier pattern helps in identifying the nodes or devices and the selection of an identifier pattern can be made according to the requirements of an application. IoT is based on interaction among the users and things. Therefore both users and things have to be identified in an IoT system. Some other identifiers required in an IoT system are service identifier, data identifier, and communication identifier.[13]

The thing identifier is a healthcare IoT system is needed to identify the things, that is, the wearables that are used to monitor a health parameter like blood pressure, glucose level, oxygen saturation, etc. The purpose of service identifiers is to identify the software applications and services and identifying the methods required for interacting with the application or service. The communication identifiers help in identifying the source and destinations, that is, the communication end points in a system. The user identifier helps in identifying the users of an IoT application or service. In an IoT system for healthcare the users are generally human beings. A user identifier system will help the system to check if the user who is accessing the given system is an authenticated user by using identification techniques like username or fingerprints. The healthcare system generates a large amount of data as it is used for continuous monitoring of the patients. The data identifier will help in identifying the specific data instances and data types. Identification of data types will help the applications to access the data from the data set for knowing the changes in health and predicting any emergencies that may arise.

The identifiers used should be unique to enable identifying of a particular entity. The identifiers used for different entities should be unique and at the same time be able to preserve the security and privacy as a lot of personal data is shared in an IoT-based healthcare system. Coding technologies like QR codes and RFID or specific pattern formats like IEEE 64-bit Extended Unique identifier can be used for generating an identifier pattern. The pattern should be short enough to cope with the constrained environment of the IoT nodes like limited battery life, processing power, and storage space.

9.8.2 COMMUNICATION TECHNOLOGY

The communication technologies in the healthcare IoT system can be broadly classified as medium range and short range. The range of communication many vary from few centimeters to several meters. The short-range communication technologies are related to the communication for a body area network (BAN) or a wearable. The long-range communication is related to the communication between the base station and BAN that is required to communicate the health parameters to the physicians and generate alerts. Short-range communications can be implemented using Zigbee, Bluetooth, RFID, and Wi-Fi. The long-range communications can be done by the cellular networks and low-power wide-area networks (LPWANs). Satellite communication can be used for the applications where the users are located at remote and widely separated geographical areas like oceans and mountain peaks as it is not possible to use cellular networks at these locations. The satellites receive signals from the Earth, amplify them and resend them to the Earth. Satellite communication provides high rates of data transfer and stability as compared to other communication technologies, but its power consumption is very high.

9.8.3 LOCATION TECHNOLOGY

The number of IoT users is increasing exponentially and similarly the need for obtaining precise data of the location of things or IoT nodes is increasing. Integration of location technologies with the IoT applications will open up new opportunities for the growth of IoT geo-location use cases. The selection of a location technology is done on the basis of whether the device has to be used indoor or outdoor and the frequency of checking the location of the device. One should know the exact duration after which the updates for an IoT device should be done like after a day, after a week, or after every minute because every update drains the battery. As all the IoT devices are battery operated, the methods used for its update should be such that it helps in maintaining long life span for the device. Some popular options for location-based service are Wi-Fi, Li-Fi, and Bluetooth Low Energy (BLE), near field communication (NFC), and Global Positioning System (GPS). GPS is preferred for outdoor communications while RFID and ZigBee can be used for short distance. Selection of a location technology for an IoT application is difficult and it has to be decided after evaluating the pros and cons of all the available techniques. A single mechanism is not sufficient to form an effective IoT system because of which a combination of the location technologies is used to create a complete solution.

In IoT-based healthcare system, location services are need for detecting the positions of ambulance, healthcare providers, patients, and caretakers. An example of locating ambulances has been discussed in Ref. [14] that uses IoT-based services to update the health parameters of a patient to the hospital. Generally, an ambulance will pick up a patient and take him to the hospital after which the doctors examine him and suggest medical treatments. A smart ambulance service will update the health parameters of the patient for the hospital staff and use a GPRS system through cloud for knowing the exact distance of the ambulance and hospital. This approach has an additional feature of turning red signals to green when the ambulance reaches within a range of 100 m. These features of location of an ambulance helps the hospital staff to make preparations for the patient's treatment before he reaches the hospital and at the same time solves the issue of traffic jams that cause delay in ambulance services. Different technologies being used in IoT healthcare are shown in Figure 9.5.

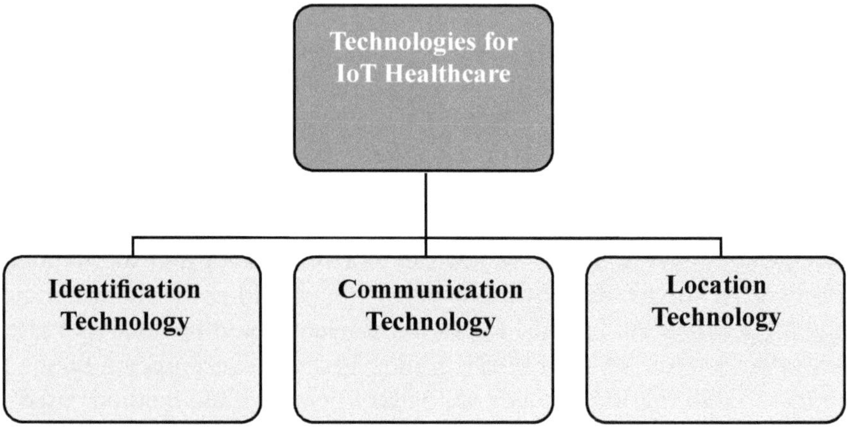

FIGURE 9.5 Technologies in IoT healthcare.

Source: Adapted from Ref. [1]

9.9 SERVICES OF HEALTHCARE IoT

The combination of IoT in the field of Healthcare has helped in devising many solutions for the Healthcare problems that have existed for long. With the rise in Healthcare demands and development of technology, the number of IoT healthcare services is increasing. The services and protocols generally used for the IoT environment need slight modification while using it for healthcare services.[12] The services in the Healthcare IoT provide a set of healthcare solutions.

9.9.1 AMBIENT-ASSISTED LIVING (AAL)

The developments in the medical field have increased life span of people which has increased the number of elderly people. The needs for looking after the elderly, providing them with physical and mental support, taking care of the health, and at the same time encouraging them for an independent living is becoming a challenge. AAL is a concept that uses the tools of information and communication technologies to improve quality of life and ensure safety and Healthcare for the older persons. As IoT started emerging as a new paradigm for healthcare services, the researchers in Ref. [15] came up with an IoT-based AAL system for healthcare. The AAL system is a four-layered framework with sensing layer, network layer, data processing layer, and application layer.

The sensor layer consists of many types of sensors. It can be in the form of a wearable or wrist band that would help to keep a track of health information like heartbeat, blood pressure, oxygen level, fall detection, etc. The other type of sensors that can be used is the user mount accelerometers that help in recognizing the daily activities like sitting, walking, brushing, relaxing, watching TV, etc. In addition to this there are ambient sensors that help to keep track of outsiders, temperature, light and water at home. The network layer collects the sensor data and with the help of wireless communication technologies sends it to the data processing layer. The data processing layer performs analysis of the data for identification of health information and other living conditions. The application layer makes use of all the information available and presents it to the user in the form of messages, notifications, graphs, line charts, etc. depending on the type of information to be conveyed to the users. This kind of intelligent health services have to deal with some challenges like storing, processing, or analyzing of the voluminous data generated by the sensors or devices in the network. The presence of many as well as different types of sensors results in an uncontrolled network where it becomes difficult to detect the new devices in the network or to know the state of device failure in the network due to power consumption.

9.9.2 ADVERSE DRUG REACTION

An adverse drug reaction (ADR) can be defined as an unintended and unpleasant reaction occurring due to the use of a medicinal product. ADR's if not attended properly can have intense affects on the quality of a patient's life and create an increased burden on the healthcare system. It does not depend on the type of disease or a medicine but varies from a person to person. The instances of ADR tend to increase with age and give rise to possibilities of drug interactions, toxicity, and unfavorable affects. In order to have a strict follow-up of drug usage and to prevent the complications arising due to incorrect drug usage, a system based on IoT has been developed in Ref. [16] for identifying drugs and monitoring medication of patients.

In this system IoT is used to inspect medicines required for treatment, detecting unfavorable side effects of pharmaceutical excipients, allergies, and side effects among pregnant women. It checks the drug, monitors treatment for drug compliance, and detects ADR. The use of IoT adds a new dimension to the task of preventing ADR by enabling the usage of application by anyone at any place. It uses RFID and NFC for drug identification which can be used with IPV6 connectivity over low-power area networks (LoWPAN) and integrated to personal devices like smart phones, tablets, and computers. All these technologies together enable the identification of devices, locate them, and connect people with these smart devices. After the identification of drug, the profile of patient is verified through the embedded knowledge-based system in the device of the user. This system checks the personal health records of the patient to have knowledge of medical history of the patient and any allergy profile. Based on these records, the system decides the suitability of a given drug for the patient. Each time a new drug is prescribed by the doctor, the system adds it to the patient history checks it and warns the doctor about the reactions of the drug. Verification of drug before consumption is important for people with chronic diseases and for the elderly. The verification is done with the help of smart phones that alert the patient about the time to consume a drug along with the assurance that the drug is safe.

9.9.3 CHILD HEALTH INFORMATION

The concept of child health information (CHI) aims to create awareness to ensure good health for children. One of the primary objectives of CHI is to empower the children as well as parents about overall health that includes nutrition, emotional, and mental behavior. Portable medical devices when enabled with IoT sensors can help in gathering medical or health information of children and notify the doctors as well as parents in case of emergencies.

The authors in Ref. [17] has proposed a framework that protects children against diseases, tracks their health records for promoting the knowledge of nutrition through edutainment and mHealth features. This healthcare system includes three types of participants: children (age 9–15), parents, and health instructors. The physical activity, heart rate, and location of the child is monitored through a smart watch. Food detection and security features are implemented through mobile platforms and are tracked by using a smart phone sensor as a camera. The system uses open source operating system AsteroidOS that provides base for the developers to build personal apps for the smart watch. The children are engaged in a simple and entertaining platform that involves gaming activities. The children answer questionnaires after which a score card is generated depending upon their performances and food choices. The children score to cross different levels in the game and simultaneously their score card gets updated. The health tutors and parents are notified about their score card after each level. The food detection functionality has been implemented using You Only Look Once (YOLO) algorithm, which is capable of providing nutrition information to the users. It is able to label the food with timing categories like breakfast, lunch, and dinner. Each time the child logins for the game, the phone camera recognizes the child by comparing it with the image stored in the memory. This step ensures security and also assists in calculating the attendance rate. All the messages, and different types of data associated with the application like videos or images are secured through end-to-end encryption. The application calculates body mass index (BMI) by allowing the user to enter height, weight, and age. Similarly, it calculates the total daily energy expenditure (TDEE) by entering their activities, which helps to know the amount of calories burnt during activities. The parents and health instructors receive notification of the score cards. The parents can also set safe zone over Google Map for their children so that they receive alerts when the child enters or leaves it. An architecture of IoT-based child health information system is shown in Figure 9.6.

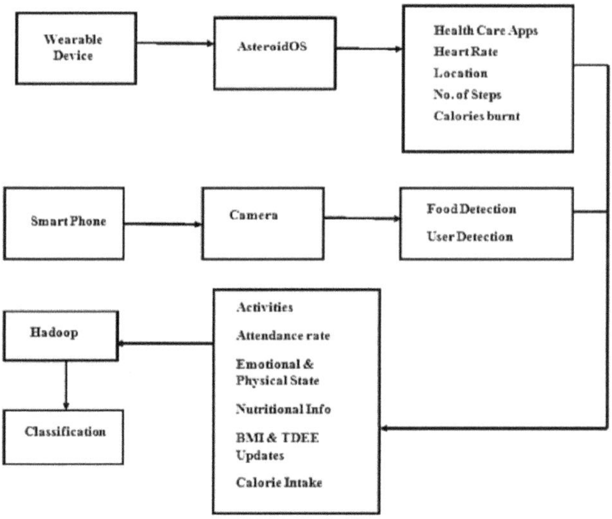

FIGURE 9.6 IoT-based child health information system.

Source: Adapted from Ref. [17]

9.9.4 COMMUNITY HEALTHCARE

The concept of community Healthcare provides support to a group of population with limited access to the health facilities especially the rural areas and remote locations. An IoT-based system can be implemented for such places where the health of people can be monitored by the team of doctors in nearby town or city. A similar concept called as IoT-equipped hospital has been explained in Ref. [18] that provides medical help to the people in rural areas.

This framework is used for the villages that lack medical facilities. It uses a bottom-up approach with three levels: village level, block level, and state level. The block level consists of health centre that aids minimum 10 villages. Any patient in the village is treated with the help of a Swasthya Slate. The Swasthya Slate is a mobile platform that can conduct 33 diagnostic tests within few hours. It is a handy tool for the health workers in rural areas where the people are either isolated due to the geographical location or are very poor to reach the nearest health centre. It provides instant results thereby facilitating better and speedy treatment. The doctor collects the patient's medical history, stores it in a smart card, and maintains a database. If the doctor at the health centre is unable to treat the patient, he seeks advice of the doctor at the state level through video conferencing. If the patient is diagnosed for a disease that cannot be treated at the health centre, the patient is transferred to the IoT-equipped hospital at the state level by a telemedicine ambulance. The telemedicine ambulance is equipped with basic medical treatment to be provided to the patient on the way to the hospital. When the patient reaches the hospital, the doctors are prepared for the treatment as they already have the patient's medical history and reports that were uploaded by the health centre in the cloud. The medical records of the patient can be accessed though the smart card issued at the health centre. The IoT-equipped hospital is equipped with a number of sensors that have been placed at different locations for management. The different types of management in an IoT-equipped hospital includes: environment management to provide data related to humidity, light, temperature, air quality, etc., water management to ensure proper distribution of water and avoid leakages, garbage monitoring to collect the garbage from rooms when the dustbin is full. Some other features of this type of hospital are telemedicine ambulance, smart wheel chair to track the movement of the patient in hospital premises, smart stretchers that are provided with sensors to know that condition of patient so that doctor can immediately start the treatment eliminating the need for time required by the doctor to examine patients on their arrival. The security of patients is maintained by providing RFID cards to the doctors and hospital staff, which is scanned by the RFID readers at the entrance of room. Health of the admitted patients is monitored through the smart beds and smart washable clothing that generates an alert to the doctors in case of emergencies. The IoT-equipped hospital can be a boon to the people as it can provide medical assistance to the people of villages, at anytime of the day with the help of internet connectivity and cloud services.

9.10 BENEFITS OF HEALTHCARE IoT

IoT has given a new dimension to the existing healthcare system. The apps and devices have taken the medical care to a new level with the help of automation. The healthy lifestyle given by the IoT has a number of benefits and few challenges too that need attention to get maximum benefits of the smart healthcare system. The challenges can be overcome to increase the adoptability of the technology among the common people. Research activities should be widened to deal with the problems faced during implementation of IoT-based healthcare systems. Some of the benefits of using IoT for Healthcare as described by the authors in Ref. [19] are remote monitoring, better clinical monitoring, and reduced errors and cost.

9.10.1 REMOTE MONITORING

Remote monitoring of the patients is one of the most discussed benefits of the smart healthcare system. The elderly, disabled, or the people with severe health problems are unable to travel to the hospital to meet the doctor for physical examination, undergo various tests, and discuss about the status of their health. Similarly people staying in geographical locations like deep forests, hilly regions, or small villages with insufficient access to transportation find it difficult to visit doctors frequently. In some cases the people are travelling for a holiday or a business trip but want to be in regular contact with the doctor or the patient needs the help of his family doctor but the doctor is not in town. In all such circumstances the smart healthcare system is a boon where the doctors are able to monitor the vital health parameters of patients without any physical contact with the patients. The patients are monitored in the comfort of their own homes and do not require hospital stays.

9.10.2 CLINICAL MONITORING

The sensors, devices, or the wearable collect the health data like blood pressure, oxygen saturation, heartbeat, etc., send the sensed data to the cloud where the data is stored and analyzed after which it is send to the devices of the doctors and the users. The doctor is thus able to know the current status of the health of his patients eliminating the need to examine them, maintain their medical records, and study them for providing them with medical assistance. The task of clinical monitoring of the patients thus becomes easier for the doctor.

9.10.3 TIME SAVING

Another advantage of the IoT healthcare system is that it saves time of both the patients as well as doctors. The patients don't have to travel for regular visit to the doctor and the diagnostic centers for checking blood pressure, glucose levels, etc. On the other hand, as the doctor can monitor the health of the patient through any device like smart phone or computer, the time required by the doctor to examine the patient is reduced. Another advantage is the economical benefits to the people as they don't have to regularly visit the doctor and do not have to pay for the medical tests at regular intervals as their health parameters are being monitored through the IoT devices.

9.10.4 REDUCED ERRORS

The continuous monitoring system of IoT Healthcare reduces the chances of errors that may result due to the manual analysis done by the medical staff according to the medical history of the patient. This system creates an environment that is needed by every patient where the improved accuracy of the diagnosis provides better experience. The continuous monitoring of the health through the sensors helps to detect any abnormalities or emergency conditions that may arise in future, generate alerts to the concerned persons which can prevent major injuries or damages, and save life of people in event of medical emergencies like heart failure, asthma attacks, etc. The continuous collection of health data is a resource for carrying research activities in the field of medical Healthcare. The data can be utilized to know the changes in health of a person according to age, gender, physical activity, and diet plans that can be used for research for developing better medicines and testing facilities. The cloud platform provides an unlimited storage to the voluminous data collected by the IoT devices which saves the money to be invested in maintain storage devices for the data. In addition to this, the faster analysis done by the cloud saves time that would be required for manual analysis using the traditional diagnostic methods.

9.10.5 AUTOMATIC UPDATE

The smart cards used for maintaining the data of the patient adds to the advantages of IoT-based Healthcare. It holds encrypted data of the patient, reduces the need of paper work preserving patient's records. It uses a digital signature or some biometric template to lessen the ambiguity of the identity of the card holder. The smart card gets updated automatically which reduces the efforts required by the healthcare workers to regularly update the information of patients in their record. The doctors can access the smart card for the health record and lab results. The digital storage process of the smart card reduces cost and time required for maintaining of data, reduces the human errors as well. Not only the doctors but the employers and the insurance agents are able to track the health of people associated with them as the health parameters get updated automatically after each doctor visit or lab tests. The insurance agencies can use this data for underwriting and claims operation. The IoT devices help to ensure transparency among the customers and insurance companies to complete the process of pricing, claim handling, and risk assessment.

9.11 CHALLENGES AND SOLUTIONS FOR IoT HEALTHCARE

Although IoT healthcare has a lot of benefits, it is highly efficient but it has its own set of challenges that hinder the widespread adoption of the technology. The IoT Healthcare can be more stable and flexible if the challenges are addressed efficiently. Some of the challenges mentioned in Refs. [20] and [21] are managing voluminous data, security of data, and authentication, interoperability while Ref. [22] discusses about the issues like energy efficiency and latency.

9.11.1 LARGE AMOUNT OF DATA

The IoT healthcare involves collection of health data sending it to the cloud and generating alerts for the users. As the number of devices and users of the technology is growing, the amount of data generated with the Healthcare application is increasing proportionally. The voluminous data from the Healthcare IoT has to undergo four types of process: collection of health data, conversion of data from analog to digital for further processing, storing digitized data in cloud, and finally analyzing the data to help the physicians in discovering treatment options for disease management. All these processes call for the implementation of data analytics and data management solutions. The smart phones used for the Healthcare applications generate unstructured data which is difficult to process as compared to the structured data. The organizations and the service providers are switching to natural language processing and artificial techniques to structure the collected data so that it can be properly utilized for the benefit of people.

9.11.2 SECURITY AND PRIVACY OF DATA

The healthcare IoT systems are connected to the internet and contain personal data which increases the vulnerability to security attacks. The data collected by the sensors has to be shared by different Healthcare units for its analysis. In this case all the Healthcare units have to ensure privacy of data. The health data is stored in cloud due to limited storage capacities of the IoT data. The limitations of IoT devices like limited bandwidth, battery life, and low memory space increase the risks of attacks like denial-of-service (DoS), man in the middle, and eavesdropping. The health data is sensitive as it is personal data. The hackers can steal the data, damage it, or manipulate it. The loss of data can be a serious problem for the Healthcare application. At the same time manipulation of the data by the attackers would result into wrong decisions by the doctor. In view of the manipulated data, the doctor may prescribe wrong medicines or treatment to the patients which may prove to be fatal. Taking into account the importance of health data, the organizations dealing with the Healthcare applications should develop and implement policy, standard, and guidelines to ensure security and privacy. The employees should be sensitized about the importance of security of health data. They should be informed about the security policies and guidelines framed by the organizations by arranging training sessions. These polices should be regularly updated as per the needs of the Healthcare sector.

9.11.3 AUTHENTICATION

The websites use passwords to verify the authentication of its users and the browser uses secure sockets layer protocol for authenticating the websites. As the people are not aware of importance of keeping strong passwords and changing them after a period, passwords alone are not sufficient for internet authentication. Single-step authentication using passwords can be worse for the IoT systems due to the different levels involved in the IoT process where the data has to pass through levels like network, cloud, etc., and undergo processing, conversion, and analysis.

In IoT healthcare system, the medical devices employed to collect the health data must authenticate to the local gateway while transferring the data. The gateway has to authenticate with the cloud end point before forwarding the data. The applications that analyze and furnish health data must authenticate with the cloud to request data. Due to the presence of multiple levels, authentication by using security tokens is required at all the levels where the token is used to identify the actor at previous level thereby enabling the present actor to make relevant authorized decision.[23] The users should be able to control the mechanism of collection of health data, the security tokens issued at different levels of interaction should not be issued without the user's consent. OAuth 2.0 and OpenID Connect 1.0 are two standardized frameworks that enable the user to involve in the process of issuance of token to the applications that access the user data for ensuring privacy control. As OAuth and OpenID Connect have been bound to HTTP only and as HTTP is insufficient for many IoT interactions, a new class of protocols MQ Telemetry Transport and Constrained Application Protocol have emerged. Using a two-factor authentication for a smart phone is considered to be a powerful authentication as most of the users carry their phones with them all the time. Similarly, the wearables used to access health data, are bond to a user that can facilitate the authentication of users when they access devices, applications, and cloud services.

9.11.4 INTEROPERABILITY

The Healthcare data are collected from different types of devices which may be in different forms like audio files, tests results, data from the wearable, patient medical history, appointment schedules, etc. The data collected by the IoT platform is heterogeneous in nature. The healthcare organizations use this data for medical research and face difficulties in managing the huge data collected by the devices. To foster research and exchange of data among the organizations interoperability is considered as an essential requirement to deal with data heterogeneity. The medical data is not only heterogeneous but is of different levels of quality. Though the data may be interoperable, transformation of data into a common format is not sufficient. The medical data plays a crucial role in the decision-making for its users; therefore, after knowing about the different quality of data available it would be able to transform the data only of high quality level. Therefore, while dealing with the medical data, it is important to deal with the heterogeneous data that is expected to be interoperable and at the same time has to identify the quality of the enormous data received. Therefore, along with data interoperability the researchers have to also deal with the data quality which is difficult to achieve in a single system most of the times.

9.11.5 ENERGY EFFICIENCY

The sensors used in the IoT devices are mostly battery-powered. Power is required by the devices to gather process and transfer data. Fast draining of battery may result in failure of nodes that will affect the communication in the healthcare system which can be disastrous in emergency situations. The failure of a node disrupts the whole network which increases the possibilities of failure in reception or transfer of the Healthcare data to the users and the physicians on proper time. The Healthcare application processes a huge amount of data in real time which increases its requirement of data storage and processing capabilities and thus contributes to consumption of energy. The cloud data centers are located far away due to which the data collected by the sensors has to be sent through multi-hop communications. The multi-hop communications and security algorithms result in more consumption of energy. Therefore there is a need to develop frameworks or algorithms that will reduce the delay caused due to multi-hop communications and provide security to the Healthcare applications without affecting the parameters like energy consumption, throughput, and delay. Fog computing when combined with cloud-based IoT system can reduce the need for multi-hop data communication and reduce the latency by allowing data processing closer to the edge of network. These capabilities of fog computing will decrease the energy consumption thereby increasing the life time of IoT nodes so that they are able to transfer data for a longer period allowing processing of real-time data for the Healthcare applications. Using artificial algorithms for cloud computing can also be helpful in making the IoT systems more energy efficient.

9.11.6 LATENCY

The latency in IoT Healthcare applications is generally higher due to the large amount of data present on the networks. As the number of users of the Healthcare application is increasing, the amount of data is increasing that increases the traffic, causes congestion in the network, and increases the latency. The cloud is used in IoT applications for the analysis and filtering of data. Due to the presence of large amount of data and multi-hop communications between the IoT devices and cloud, it becomes difficult for the cloud to fulfill the minimum latency requirements of the Healthcare applications. Another issue is that as the healthcare relies on real-time data, high latency will make the data meaningless and inadequate for its users. The physiological data of a patient keeps on varying with time. Monitoring of remote patients can be done successfully if the data is received by the doctors on time to facilitate rapid decisions during emergencies. A delay of microsecond or millisecond can be crucial for a Healthcare application. Using fog computing layer between the IoT devices and cloud computing layer will reduce the burden on cloud and as the data can be processed in the fog layer, the overall latency can be reduced. Machine learning and deep algorithms when combined with the fog computing approach can yield better results.

9.12 CONCLUSION

Using IoT for Healthcare has made it easier for the health professionals to monitor the health of their patients. The sensors collect the physiological data, send it to the cloud for processing after which it is send to the doctors, and also alerts are generated if any emergency conditions are detected. In traditional methods, the doctor starts the treatment only after examining the patient which may be time consuming. IoT-based Healthcare can be a boon in emergencies as it is able to detect the abnormalities and sends notifications after which the doctor can immediately response to suggest medical care to save the life of patients. It helps in remote monitoring, which saves the time in visiting the doctor. It is beneficial especially for the elderly and people with disabilities who cannot travel long distances. Healthcare IoT has some challenges too like security and privacy, authentication, and huge amount of data. Though it is difficult to develop a holistic approach that will be able to solve all the challenges but approaches like fog computing and artificial intelligence can be used to reduce the effects of these challenges. Dealing with challenges is important as any negligence in the Healthcare applications can be dangerous for its users. The field of Healthcare can be explored more to improve the experience of the people and improve its adoptability among the people.

KEYWORDS

- **IoT**
- **healthcare**
- **security**
- **medical devices**

REFERENCES

1. Pradhan, B.; Bhattacharyya, S.; Pal, K. IoT-Based Applications in Healthcare Devices. *J. Healthc. Eng.* **2021,** *2021*, Article ID 6632599, https://doi.org/10.1155/2021/6632599

2. Jia, T. N.; Ali, M.; Dhaou, I. B.; Rahmani, A. M.; Westerlund, T.; Liljeberg, P. In *IoT-Based Continuous Glucose Monitoring System: A Feasibility Study*, International Conference on Ambient Systems, Network and Technologies (ANT-2017) Procedia, https://doi.org/10.1016/j.procs.2017.05.359
3. Shaown, T.; Hasan, I.; Mim, M. M. R.; Hossain, M. S. In *IoT-Based Portable ECG Monitoring System for Smart Healthcare, 2019*, 1st International Conference on Advances in Science, Engineering and Robotics Technology (ICASERT), 2019; pp 1–5. DOI: 10.1109/ICASERT.2019.8934622.
4. Hashim, N.; Norddin, N.; Idris, F.; Yusoff, S. N. I. M.; Zahari, M. IoT Blood Pressure Monitoring System. *Indonesian J. Electr. Eng. Comput. Sci.* **2020,** *19*(3), 1384–1390. http://doi.org/10.11591/ijeecs.v19.i3.
5. Hoang, M. L.; Carratù, M.; Paciello, V.; Pietrosanto, A. Body Temperature—Indoor Condition Monitor and Activity Recognition by MEMS Accelerometer Based on IoT-Alert System for People in Quarantine Due to COVID-19. *Sensors* **2021,** *21*, 2313. https://doi.org/10.3390/s21072313
6. Tariq, Z.; Shah, S. K.; Lee, Y. In *Speech Emotion Detection Using IoT Based Deep Learning for Healthcare, 2019*, IEEE International Conference on Big Data (Big Data), 2019; pp 4191–4196. DOI: 10.1109/BigData47090.2019.9005638.
7. Mano, L. Y.; Faiçal, B. S.; Nakamura, L. H. V.; Gomes, P. H.; Libralon, G. L.; Meneguete, R. I.; Ueyama, J. Exploiting IoT Technologies for Enhancing Health Smart Homes Through Patient Identification and Emotion Recognition. *Comput. Commun.* **2016,** *89–90*, 178–190. DOI: 10.1016/j.comcom.2016.03.010.
8. Kurita, T.; Matsuo, K.; Barolli, L. A Wheelchair Management System Using IoT Sensors and Agile-Kanban. In *Advances in Intelligent Networking and Collaborative Systems. INCoS 2019. Advances in Intelligent Systems and Computing*; Barolli, L., Nishino, H., Miwa, H., Eds.; Springer: Cham, 2020; Vol. 1035. https://doi.org/10.1007/978-3-030-29035-1_9
9. Yacchirema, D.; de Puga, J. S.; Palau, C.; Esteve, M. Fall Detection System for Elderly People Using IoT and Big Data. *Proc. Comput. Sci.* **2018,** *130*(9) 603–610. URL: http://www.sciencedirect.com/science/article/pii/S1877050918304721, the 9th International Conference on Ambient Systems, Networks and Technologies (ANT 2018)/The 8th International Conference on Sustainable Energy Information Technology (SEIT-2018)/Affiliated Workshops. https://doi.org/10.1016/j.procs.2018.04.110.
10. Mrozek, D.; Koczur, A.; Małysiak-Mrozek, B. *Fall Detection in Older Adults with Mobile IoT Devices and Machine Learning in the Cloud and on the Edge. Information Sciences*, 2020; Vol. 537, pp 132–147. https://doi.org/10.1016/j.ins.2020.05.070.
11. Miron-Alexe, V. IoT Pulse Oximetry Status Monitoring for Home Quarantined Covid-19 Patients. *J. Sci. Arts* **2020,** *20*(4), 1029–1036. http://dx.doi.org/10.46939/J.Sci.Arts-20.4-c04
12. Riazul Islam, S. M.; Kwak, D.; Kabir, Md. H.; Hossain, M.; Kwak, K. S. The Internet of Things for Healthcare: A Comprehensive Survey. *IEEE Access* **2015,** *3*, 678–708. https://doi.org/10.1109/ACCESS.2015.2437951
13. Identifiers in Internet of Things, Alliance of Internet of Things Innovation, Version 1.0, February 2018.

14. O. Udawant; N. Thombare; D. Chauhan; A. Hadke; D. Waghole. In *Smart Ambulance System Using IoT*, 2017 International Conference on Big Data, IoT and Data Science (BID), 2017; pp 171–176. DOI: 10.1109/BID.2017.8336593.
15. Wan J.; X. Gu; L. Chen; J. Wang. In *Internet of Things for Ambient Assisted Living: Challenges and Future Opportunities,* 2017 International Conference on Cyber-Enabled Distributed Computing and Knowledge Discovery (CyberC), 2017; pp 354–357. DOI: 10.1109/CyberC.2017.83.
16. Jara, A. J.; Miguel A. Zamora; Antonio F. Skarmeta. Drug Identification and Interaction Checker Based on IoT to Minimize Adverse Drug Reactions and Improve Drug Compliance. *Pers. Ubiquit. Comput.* **2012.** http://dx.doi.org/10.1007/s00779-012-0622-2.
17. Nishargo Nigar, Mohammed nazim uddin. In *An Intelligent Children Healthcare System in the Context of Internet of Things*, 2nd International Conference on Innovative Research in Science, Technology & Management, National University of Singapore (NUS), June, 2018.
18. Kakkar, A.; Shaurya. An IoT Equipped Hospital Model: A New Approach for E-governance Healthcare Framework. *Int. J. Med. Res. Health Sci.* **2019,** *8*(3), 36–42. ISSN No: 2319-5886.
19. Mohan, T. Y; Raja, V. Internet of Things: Benefits and Risk of Smart Healthcare Application. *J. Inform. Syst. Res. Innov.* **2016,** *10*(3), 37–42.
20. How IoT is Shaping the Future of the Healthcare Industry, Whitepaper. www.zerone-consulting.com
21. Sherali Zeadally; Farhan Siddiqui; Zubair Baig; Ahmed Ibrahim. *Smart Healthcare, Challenges and Potential Solutions Using Internet of Things (IoT) and Big Data Analytics*; Emerald Publishing Limited, 2020; Vol. 4, No. 2, pp 93–109, http://dx.doi.org/10.1108/PRR-08-2019-0027
22. Naresh, V. S.; Pericherla, S. S.; Rama Murty, P. S.; Reddi, S. Internet of Things in Healthcare: Architecture, Applications, Challenges, and Solutions. *Int. J. Comput. Syst. Sci. Eng.* **2020,** *6*, 411–421. http://dx.doi.org/10.32604/csse.2020.35.411.
23. https://www.secureidnews.com/news-item/authentication-in-the-iot-challenges-and-opportunities/

CHAPTER 10

APPLICATION OF IoT IN CANCER

ARCHANA SINGH[1], WASEEM RAZA[1], and SHASHIKANT VERMA[2]

[1]*JK Cancer Institute, Kanpur, Uttar Pradesh, India*

[2]*G.B. Pant Institute of Engineering & Technology, Pauri Garhwal, Uttarakhand, India*

ABSTRACT

Cancer is the word that describes whole group of diseases that are caused by uncontrolled multiplication of any type of cell in the body. No single reason can be the cause of cell division going uncontrolled, multiple reasons in a single or combined way lead to the initiation and development of cancer. Cancer can affect any part of the body initially but can spread locally, distant ultimately leading to death if not treated accordingly. Cancer is heterogeneous with number of histological and biological variety. Each of them requires specific strategies for diagnosis and treatment which in turn require qualified and coordinated teamwork for multidisciplinary patient care.

10.1 INTRODUCTION

Approximately 19 million new cancer cases have already been documented with 50% mortality rate. Out of these cancer cases 50% were from Asia where 60% of the global population resides.[1,2]Lung cancer and breast cancer are the most frequently occurring cancers among male and female, respectively, also the leading cause of cancer deaths. Such huge rate of

IoT-Enabled Healthcare Systems: Applications, Benefits, and Challenges
Yatindra Kumar, Shashi Kant Verma, Archana Singh, Krishna Kumar, & Manoj Gupta (Eds.)

morbidity and mortality causes massive economic burden not only to the individual or family but country and globally too.[3] One out of six deaths is contributed by cancer. Annually more than 9 million people are dying due to cancer. The majority of these are from low-income and middle-income countries. Due to all given reasons cancer prevention and control mandates prioritization by both national and international level in the form of investment and infrastructure. Cancer incidence is estimated to have doubled by2035.Therefore, each and every country requires its own robust infrastructure for not only screening, diagnosing, and treating but also creating future plans related to cancer. So accurate assessment of the problem in such diversely populated countries necessitate use of internet of things (IoT), artificial intelligence, biosensors, cloud computing overall technology-assisted system for collecting our own data making it useful for screening, diagnosis treatment, follow-up, general care, research, prevention, etc.

10.1.1 INTERNET OF THINGS (IoT)

IoT word is coined by Mr. Kevin Ashton in association with supply chain a decade ago.[4] Computing and internet-related technical advances lead to use of IoT in healthcare. Sensors, wearable devices, appliances, controller, etc. are the smart objects connected with internet called IoT used not only for treating disease but also for prevention to save overall health ecosystem. So to conquer cancer is not just a disease and treatment, it's a systematic process to know the burden then deal with by using all resources. Disproportion exists among and across the countries in cancer presentation diagnosis treatment, etc. Accordingly, patient-oriented health system to mark improvement in cancer care is the need of this era. Therefore, technology-driven solutions are needed to speed up cancer prevention and treatment which in turn decrease mortality drastically. To achieve these set of goals requires technical training and robust health information system to deliver cancer services timely.

10.1.2 IoT DEVICES AND CANCER

Smartphone applications improve patients' motivation and clinical efficacy in many chronic diseases, leads to technical development of

IoT-based devices in healthcare especially cancer. In a randomized controlled trial by Galiano-Castillo et al. in breast cancer patients showed better strength of the body post anticancer treatment by using IoT devices may be due to more organized follow-up plans assessed in the form of handgrip strength.[5] Assessed pulmonary rehabilitation by using smart phones in the lung cancer cases post chemotherapy and found better exercise capacity with decrease psychological issues in the form of anxiety, depression.[6] Most cancer patients require rehabilitation in one and other form posttreatment, but various obstacles like distant location of available services, time-consuming procedures, also high costs lead to decreased quality of life. Smart after care with the use of available IoT devices improves quality of life efficiently, overcoming all obstacles.[7]IoT plays a major role in interoperability in different departments, different specializations, data monitoring and maintaining issues, etc as required in cancer management.[8]IoT-based data-centric processes are associated with many challenges like security, privacy, integration, etc. Processing and storage of redundant data can be damaging in health-related events especially cancer. Hence, raw telemedicine data can be used effectively by machine learning and deep learning techniques. Various tumor markers are already known to have a pivotal role in diagnosing cancers, but by the help biomarker-based detection strategies make cancer detection early and easy.[9] Artificial intelligence-based technologies plays important role in each and every area of cancer screening, research, diagnosis, prediction of patient outcomes, treatment responses, personalized treatment, automated radiotherapy workflow, and novel anticancer drug discovery, etc. Various prediction models by use of AI and ML are used to know lymph nodes metastasis, treatment response, and prognosis. Approximately, 82% of accuracy has been reported in detecting gastric cancer preoperatively by using clinical data, pathological data, and genetic polymorphisms.[10] Gene expression is an efficient method but associated with limited sample sizes. The few used prediction algorithm are K-nearest neighbor (KNN), support vector machine (SVM), and decision tree (DT). AI-aided (DAB)-based immunohistochemical test detected HER2 overexpression in breast cancer cases with sensitivity and specificity of 95–100%.[11] Table 10.1 shows the applications of biosensors in cancer studies.

TABLE10.1 BIOSENSORS and Cancer.

Types of biosensor	Biomarker detected	Cancer diagnosis	Studied by
CNT-based biosensor	Volatile organic compounds	Ca Lung	Liu et al. [12]
	PSA	CA prostate	Feng et al. [13]
	HeLa,HL60	Cancer	Zheng et al. [14]
	TP53	Cancer	Fayazfar et al. [15]
	HER2	Breast cancer	Arkan et al. [16]
	Cyclin a peptide	Cancer	Wang et al. [17]
Surface plasmon resonance-based biosensors	CEA	Colorectal cancer	Altintas et al. [18]
	PSA	Ca prostate	Jang et al. [19]
Quantum dot-based biosensors	miR-141	Ca prostate	Wilner et al. [20]
	TdT	Leukemia	Zhang et al. [17]

10.1.3 LUNG CANCER

Lung cancer is the most common cause of cancer deaths accounting approximately18% of total cancer death and is also second commonly diagnosed cancer. It is the 3rd most common diagnosed only after breast and colorectal malignancies in women.[18]. The 5-years survival rate of patients with lung cancer is only 10–20% in most countries. Screening with low-dose CT scan in few countries resulted in a lung cancer mortality reduction to about 24% in men and 33% in women. In early stage lung cancer especially when lung nodule is less than 1cmdeep learning-based CT scan analysis techniques do better than radiologists. The distinction between benign and malignant nodules is very difficult and challenging. Clinical biomarkers, especially plasma proteins and blood tests are very useful for the classification of early-stage lung nodules.[19]Computer-assisted decision support system developed by Masood et al. showed lung nodule detection very accurate and efficient.[20] Zhang et al.[21]showed differentiation between healthy and lung cancer by pulse signal processing method analyzing pulse waveform of wrist pulse applying SVM. So artificial intelligence, deep learning, and IoT-based methods can be very helpful in correlating, clinical, pathological, and radiological findings in screening, as well as follow-up lung cancer cases.

10.1.4 BREAST CANCER

Biopsy is a highly accurate diagnostic method in case of breast cancer, but it is invasive. A mammogram is a breast imaging technique to detect and differentiate benign and malignant lesions by a 2-dimensionalprojection image. MRI provides excellent results by creating 3D images along with dynamic functionality display but it is complex and time-consuming. Therefore, machine learning techniques were found to be more reliable and effective to resolve these complexities of both invasive and noninvasive methods. Below mentioned studies showed accuracy up to 99% in predicting, classifying, and diagnosing breast cancer by using various techniques.[9]

Researcher	Basis	Accuracy	Function
Azar, El-said [22]	Radial basis function Probabilistic neural network Multilayer perceptron	97%	
Alickovic and Subasi [23]	Feature selection algorithm Rotation forest	99%	Prediction
Hasan et al. [24]	Symbolic regression of multigene genetic programming	99%	
Akay [25]	F-score method and support vector machine		Diagnosis
Zheng et al. [26]	K-means algorithm		Classification
Usha et al. [27]	Hybridized principal component analysis		Classification
Marcano-Cedeño et al. [28]	Artificial meta plasticity multilayer perceptron	99.26%	Classification
Liu et al. [29]	under sampling technique		Prediction
Sheikhpour et al. [30]	Swarm optimization integrated with nonparametric kernel density		Prediction

10.1.5 GASTRIC CANCER

Upper GI endoscopy is the procedure to diagnose any gastric carcinoma case which is expensive, invasive as well as infrastructure driven so in Asian countries not method of choice for screening purpose as used in

Japan and Korea where it is commonly found.[31]Sensor-based technologies are the only hope to diagnose such cancer in early stages. Various sensor-based IoT using exhaled breath makes diagnosis of gastric cancer very fast, cost effective, even in remote areas due to its portable design.[32] These techniques can be performed routinely in wide range ages can detect cancerous as well as noncancerous diseases simultaneously. Various techniques are already devised with their pros and cons like breath analyzer, SniffPhone techniques, eNose, etc. Gold nanoparticles-based sensor technology is used in SniffPhone device to analyze breath samples. Most important aspect of these devices is calibration with dedicated standards to level many different systems.

10.1.6 PROSTATE CARCINOMA

Prostate cancer is the second most frequent cancer and the fifth leading cause of cancer death. It is the most frequently diagnosed cancer among western countries men. In approximately 50% of prostate cancer patients, androgen deprivation therapy (ADT) is used apart from surgery, radiotherapy, and chemotherapy. Decreased bone mineral density, weight gain, and cardiovascular risk increased leads to impaired quality of life are the side effects associated with prostate cancer treatment. These adverse effects can be reversed by exercise, diet, and nutritional interventions. These interventions require healthcare monitoring and counseling in long term. IoT-based technologies via wearable devices have the advantage of patient education, proper counseling supportive care accessibility anytime anywhere with proper efficacy, quality, and timeliness. The Smart After-Care (SAC) service coordinates services of physicians, counselors to patients by the use of IoT-based platform to detect and transmit any abnormal parameter incorporating mobile sensor networks to deliver personalized exercise and diet.[33]

10.1.7 CARCINOMA OVARY

Among all gynecological malignancy ovarian carcinoma is the leading cause of death. Most of the cases are diagnosed in advanced stages due to vague presenting symptoms. Therefore, it is crucial to detect it as early as possible to correctly identify the cancer stage for effective treatment options. Histopathological diagnosis is the only way to differentiate benign and malignant pelvic masses. Noninvasive methods like sensitive biomarker detection by IoT-based smart devices is the only way to screen and diagnose ovarian carcinoma in early stages. At present diagnosis and follow-up is based on the established biomarker glycoprotein named CA125 (carbohydrate antigen).[34] Ca 125 holds low specificity as a diagnostic test because it is found raised in various benign gynecological conditions. So over the last few years many other biomarkers evaluated especially for epithelial ovarian carcinoma. Human epididymis 4(HE4) appears promising in diagnosing even in early stage, treatment response monitoring, also in prognosis.[35] Improvement in clinical diagnosis and monitoring during treatment is seen due to immunosensor-based electrochemical devices. Few devices that directly display analytical concentration of the biomarker available commercially are 910 PSTAT and Drop Stat. Such autonomous devices with the help of IoT can make ovarian carcinoma diagnosis in early stages which in turn reduces overall mortality and morbidity.

10.1.8 COLORECTAL CARCINOMA

Prevention plays a major role in the case of colorectal cancer as it is found to be associated with dietary habits, lifestyles, and genetic syndromes.[36] So, screening, modifying diets, and lifestyles, etc. works miracles in decreasing morbidity as well as mortality. Colonoscopy holds a sensitivity and specificity of 95% for finding polyps and colon cancer. But it's an observer-dependent procedure, so computer-aided systems are incorporated to find suspicious regions. Colonoscopy missed polyps of size less than 1cm and flat lesions.[37] Chromoscopy used dye to detect lesions on the mucosa but is very labor intensive. Virtual chromoendoscopy with the use of AI by using various methods like Olympus Narrow Band Imaging, autofluorescence imaging, Pentax i-scan, etc. makes the processes easier and better. In the near future, an aging population may pose challenges in accessing healthcare services. IoT, IoMT, and wireless sensor networks, etc. can create wonders in cancer care especially preventable cancers like colorectal by monitoring health at various levels.[38,39]

10.1.9 BRAIN TUMORS

Imaging-based diagnosis is the key as biopsy being invasive and complicated procedure requires super specialty skills. Sometimes defining benign versus malignant lesions or grading brain tumor is challenging without biopsy. In brain tumors, a major challenge is a radiological diagnosis as biopsy in this case is a very complicated process requiring super specialty. Few below-mentioned studies uses IoT for differentiating benign versus malignant, also classifying brain tumors with accuracy of approximately more than 90%.[3–7]

Tumor	IoT method	Studied by	Accuracy
Benign and malignant	CNN+KNN	Srinivas, B. and Rao, G. S. [40]	96%
Glioma, meningioma, and pituitary	SE-ResNet	Ghosal, P. [41]	94%
Metastasis, meningiomas, Grade II and III, GBM	SVM, ANN	Zacharaki [42]	80%
Benign and malignant	Optimal dense convolutional network	Ganesan [43]	99%
Grade I, II, III, & IV	Deep convolutional neural network	Sajjad, M. [44]	91%

10.1.10 RADIOTHERAPY

Radiotherapy involves many steps at different levels to achieve effective patient outcome. Radiotherapy requires positioning, immobilization, simulation, tumor, and organ contouring, treatment planning, verification, and then execution. All these steps are time-consuming along with resources. Quality treatment requires every step to be perfect. So, radiotherapy workflow can be maintained by using artificial intelligence for automation and workflow optimization. Each step of radiotherapy creates huge amount of unstructured and unorganized data which can only be managed by using technology like knowledge maps, SVM, and neural network. Computing helps machines to learn language, vision, cognition, reasoning, etc. also known as machine learning or intelligence. This machine learning helps radiotherapy staff by managing medical records, target volume delineation, treatment planning, quality checks, etc. By incorporating block chain radiotherapy, computing,5G communicating network, various wearable devices radiotherapy can be automated as well as shared remotely in real time. All the above-mentioned technical innovations decreases radiotherapy work processing rapid, leading to radiation oncologists volume delineation very precise and radiotherapy dose calculations proper with the help of neural network and DVH prediction model. Apart from improving quality of the treatment, also remove errors related to human making treatment more consistent. IOT and AI-assisted Chatbot's in radiotherapy amalgamates various stations like OPD, IPD, day-care wards, treatment room for proper two-way communication, data transfer, appointment management, etc.[45–47]

10.2 CONCLUSION AND FUTURE SCOPE

Each and everything comes with their own cost, so is the case of IoT devices and system. With such huge database, data security will be a prime concern. Such sensitive health-related data can be misused. Also, the network of various IoT devices connected through same internet can lead to many access points for cyberattacks. Real-time hacking can cause severe consequences. So, data security, data privacy, data maintenance, data interoperability, data ownership, data validation, etc will be the prime challenge to handle in the case of healthcare system, especially cancer.[48]

In the near future, IoT systems will be available for patients continuous biometric monitoring leading customized case wise treatment and care. This will add data regarding patient's environment-related details during treatment as well as long-term follow-up making more clear understanding of cancer-related risk factors. This micro detailing of each and every case makes better preventive as well as treatment guidelines related to cancer.

KEYWORDS

- **IoT**
- **cancer**
- **carcinoma ovary**
- **lung cancer**
- **breast cancer**

REFERENCES

1. G. M., Cooper. The Development and Causes of Cancer, 2000. [Online]. https://www.ncbi.nlm.nih.gov/books/NBK9963/ (accessed Dec 20, 2022).
2. Sung, H.; et al. Global Cancer Statistics 2020: GLOBOCAN Estimates of Incidence and Mortality Worldwide for 36 Cancers in 185 Countries," CA. *Cancer J. Clin.* **2021,** *71*(3), 209–249. DOI: 10.3322/CAAC.21660.
3. Prager, G. W.; et al. Global Cancer Control: Responding to the Growing Burden, Rising Costs and Inequalities in Access. *ESMO Open* **2018,** *3*(2). DOI: 10.1136/esmoopen-2017-000285.
4. That 'Internet of Things' Thing - RFID Journal. https://www.rfidjournal.com/that-internet-of-things-thing (accessed Nov 24, 2022).
5. Galiano-Castillo, N.; et al. Telehealth System: A Randomized Controlled Trial Evaluating the Impact of an Internet-Based Exercise Intervention on Quality of Life, Pain, Muscle Strength, and Fatigue in Breast Cancer Survivors. *Cancer* **2016,** *122*(20) 3166–3174. DOI: 10.1002/CNCR.30172.
6. Park, S.; et al. Mobile Phone App-Based Pulmonary Rehabilitation for Chemotherapy-Treated Patients with Advanced Lung Cancer: Pilot Study. *JMIR mHealth uHealth* **2019,** *7*(2). DOI: 10.2196/11094.
7. Goode, A. D.; Lawler, S. P.; Brakenridge, C. L.; Reeves, M. M.; Eakin, E. G. Telephone, Print, and Web-Based Interventions for Physical Activity, Diet, and Weight Control Among Cancer Survivors: A Systematic Review. *J. Cancer Surviv.* **2015,** *9*(4), 660–682. DOI: 10.1007/S11764-015-0442-2.
8. Albino de Queiroz, D.; André da Costa, C.; Aparecida Isquierdo Fonseca de Queiroz, E.; Folchini da Silveira, E.; da Rosa Righi, R. Internet of Things in Active Cancer Treatment: A Systematic Review. *J. Biomed. Inform.* **2021,** *118*, 103814. DOI: 10.1016/J.JBI.2021.103814.
9. Choi, Y. E.; Kwak, J. W.; Park, J. W. Nanotechnology for Early Cancer Detection. *Sensors* **2010,** *10*, 428–455. DOI: 10.3390/s100100428.
10. Jin, P.; et al. Artificial Intelligence in Gastric Cancer: A Systematic Review. *J. Cancer Res. Clin. Oncol.* **2020,** *146*(9), 2339–2350. DOI: 10.1007/S00432-020-03304-9/METRICS.
11. Fan, L.; et al. Artificial Intelligence-Aided Multiple Tumor Detection Method Based on Immunohistochemistry-Enhanced Dark-Field Imaging. *Anal. Chem.* **2022,** *94*(2), 1037–1045. DOI: 10.1021/ACS.ANALCHEM.1C04000.
12. Ovádeková, R.; Jantová, S.; Letašiová, S.; Štepánek, I.; Labuda, J. Nanostructured Electrochemical DNA Biosensors for Detection of the Effect of Berberine on DNA From Cancer Cells. *Anal. Bioanal. Chem.* **2006,** *386*(7–8), 2055–2062. DOI: 10.1007/S00216-006-0830-6/METRICS.
13. Zeng, X.; Ma, S.; Bao, J.; Tu, W.; Dai, Z. Using Graphene-Based Plasmonic Nanocomposites to Quench Energy From Quantum Dots For Signal-on Photoelectrochemical Aptasensing. *Anal. Chem.* **2013,** *85*(24), 11720–11724. DOI: 10.1021/AC403408Y.

14. Park, Y. K.; et al. D-(+)-Galactose-Conjugated Single-Walled Carbon Nanotubes as New Chemical Probes for Electrochemical Biosensors for the Cancer Marker Galectin-3. *Int. J. Mol. Sci.* **2011,** *12*(5), 2946–2957. DOI: 10.3390/IJMS12052946.
15. Fayazfar, H.; Afshar, A.; Dolati, M.; Dolati, A. DNA Impedance Biosensor for Detection of Cancer, TP53 Gene Mutation, Based on Gold Nanoparticles/Aligned Carbon Nanotubes Modified Electrode. *Anal. Chim. Acta* **2014,** *836*, 34–44. DOI: 10.1016/J.ACA.2014.05.029.
16. Arkan, E.; Saber, R.; Karimi, Z.; Shamsipur, M. A Novel Antibody-Antigen Based Impedimetric Immunosensor for Low Level Detection of HER2 in Serum Samples of Breast Cancer Patients Via Modification of a Gold Nanoparticles Decorated Multiwall Carbon Nanotube-Ionic Liquid Electrode. *Anal. Chim. Acta* **2015,** *874*, 66–74. DOI: 10.1016/J.ACA.2015.03.022.
17. Wang, L. J.; Luo, M. L.; Zhang, Q.; Tang, B.; Zhang, C. Y. Single Quantum Dot-Based Nanosensor for Rapid and Sensitive Detection of Terminal Deoxynucleotidyl Transferase. *Chem. Commun.* **2017,** *53*(80), 11016–11019. DOI: 10.1039/C7CC05485H.
18. Altintas, Z.; Uludag, Y.; Gurbuz, Y.; Tothill, I. E. Surface Plasmon Resonance Based Immunosensor for the Detection of the Cancer Biomarker Carcinoembryonic Antigen. *Talanta* **2011,** *86*(1), 377–383. DOI: 10.1016/J.TALANTA.2011.09.031.
19. Jang, H. S.; Park, K. N.; Kang, C. D.; Kim, J. P.; Sim, S. J.; Lee, K. S. Optical Fiber SPR Biosensor With Sandwich Assay for the Detection of Prostate Specific Antigen. *Opt. Commun.* **2009,** *282*(14), 2827–2830. DOI: 10.1016/J.OPTCOM.2009.03.078.
20. Jou, A. F. J.; et al. Diagnosing the miR-141 Prostate Cancer Biomarker Using Nucleic Acid-Functionalized CdSe/ZnS QDs and Telomerase. *Chem. Sci.* **2015,** *6*(1), 659–665. DOI: 10.1039/C4SC02104E.
21. Masood, A.; et al. Computer-Assisted Decision Support System in Pulmonary Cancer Detection and Stage Classification on CT Images. *J. Biomed. Inform.* **2018,** *79*, 117–128. DOI: 10.1016/J.JBI.2018.01.005.
22. Azar, A. T.; El-Said, S. A. Probabilistic Neural Network for Breast Cancer Classification. *Neural Comput. Appl.* **2013,** *23*(6), 1737–1751. DOI: 10.1007/S00521-012-1134-8.
23. Aličković, E.; Subasi, A. Breast Cancer Diagnosis Using GA Feature Selection and Rotation Forest. *Neural Comput. Appl.* **2017,** *28*(4), 753–763. DOI: 10.1007/S00521-015-2103-9.
24. Kamrul Hasan, M.; Milon Islam, M.; Hashem, M. M. A. Mathematical Model Development to Detect Breast Cancer Using Multigene Genetic Programming.
25. Fatih Akay, M. Support Vector Machines Combined with Feature Selection for Breast Cancer Diagnosis. DOI: 10.1016/j.eswa.2008.01.009.
26. Zheng, B.; Yoon, S. W.; Lam, S. S. Breast Cancer Diagnosis Based on Feature Extraction Using a Hybrid of K-Means and Support Vector Machine Algorithms. *Expert Syst. Appl.* **2014,** *41*(4), 1476–1482. DOI: 10.1016/J.ESWA.2013.08.044.
27. Usha, K.; Sri, R.; Visvavidyalayam, P. M.; Naga, G.; Usha, R. K.; Lavanya, R. D. Importance of Feature Extraction for Classification of Breast Cancer Datasets-A Study Artificial Neural Network for Smart Environment View project Cervical Cancer Diagnosis Using Softcomputing View Project Lavanya Doddipalli Importance of Feature Extraction for Classification of Breast Cancer Datasets-a Study. *Int. J. Sci.*

Innov. Math. Res. [Online] **2015,** *3*, 763–768. www.arcjournals.org (accessed Dec 22, 2022).
28. Marcano-Cedeño, A.; Quintanilla-Domínguez, J.; Andina, D. Breast Cancer Classification Applying Artificial Metaplasticity Algorithm. *Neurocomputing* **2011,** *74*(8), 1243–1250. DOI: 10.1016/J.NEUCOM.2010.07.019.
29. Liu, Y. Q.; Wang, C.; Zhang, L. Decision Tree Based Predictive Models for Breast Cancer Survivability on Imbalanced Data. *3rd Int. Conf. Bioinforma. Biomed. Eng. iCBBE* **2009,** *2009*. DOI: 10.1109/ICBBE.2009.5162571.
30. Sheikhpour, R.; Sarram, M. A.; Sheikhpour, R. Particle Swarm Optimization for Bandwidth Determination and Feature Selection of Kernel Density Estimation Based Classifiers in Diagnosis of Breast Cancer. *Appl. Soft Comput.* **2016,** *40*, 113–131. DOI: 10.1016/J.ASOC.2015.10.005.
31. Choi, K. S.; Suh, M. Screening for Gastric Cancer: The Usefulness of Endoscopy. *Clin. Endosc.* **2014,** *47*(6), 490. DOI: 10.5946/CE.2014.47.6.490.
32. Leja, M.; et al. Sensing Gastric Cancer Via Point-of-Care Sensor Breath Analyzer. *Cancer* **2021,** *127*(8), 1286–1292. DOI: 10.1002/CNCR.33437.
33. Salem, M.; Elkaseer, A.; El-Maddah, I. A. M.; Youssef, K. Y.; Scholz, S. G.; Mohamed, H. K. Non-Invasive Data Acquisition and IoT Solution for Human Vital Signs Monitoring: Applications, Limitations and Future Prospects. *Sensors (Basel)* **2022,** *22*(17). DOI: 10.3390/S22176625.
34. van der Burg, M. E. L.; Lammes, F. B.; Verweij, J. CA125 in Ovarian Cancer. *Biomark. Med.* **2007,** *1*(4), 513. DOI: 10.2217/17520363.1.4.513.
35. Li, J.; et al. HE4 as a Biomarker for Ovarian and Endometrial Cancer Management. *Expert Rev. Mol. Diagn.* **2009,** *9*(6), 555. DOI: 10.1586/ERM.09.39.
36. Chan, A. T.; Giovannucci, E. L. Primary Prevention of Colorectal Cancer. *Gastroenterology* **2010,** *138*(6), 2029. DOI: 10.1053/J.GASTRO.2010.01.057.
37. Ahn, S. B.; Han, D. S.; Bae, J. H.; Byun, T. J.; Kim, J. P.; Eun, C. S. The Miss Rate for Colorectal Adenoma Determined by Quality-Adjusted, Back-to-Back Colonoscopies. *Gut Liver* **2012,** *6*(1), 64. DOI: 10.5009/GNL.2012.6.1.64.
38. Graham, D. G.; Banks, M. R. Advances in Upper Gastrointestinal Endoscopy. *F1000Res.* **2015,** *4*, F1000 Faculty Rev-1457. DOI: 10.12688/F1000RESEARCH.6961.1.
39. Taghiakbari, M.; Mori, Y.; von Renteln, D. Artificial Intelligence-Assisted Colonoscopy: A Review of Current State of Practice and Research. *World J. Gastroenterol.* **2021,** *27*(47), 8103. DOI: 10.3748/WJG.V27.I47.8103.
40. Sasibhushana Rao, G. A Hybrid CNN-KNN Model for MRI Brain Tumor Classification. *Artic. Int. J. Recent Technol. Eng.* **2019,** *2*, 5230. DOI: 10.35940/ijrte.B1051.078219.
41. Ghosal, P.; Nandanwar, L.; Chakraborty, J.; Kanchan, S. Brain Tumor Classification Using ResNet-101 Based Squeeze and Excitation Deep Neural Network Deep learning for screening of interstitial lung disease patterns in high-resolution CT images View project A Segmentation Approach to Delineate Salient Brain Tissues in MRI View project Brain Tumor Classification Using ResNet-101 Based Squeeze and Excitation Deep Neural Network. DOI: 10.1109/ICACCP.2019.8882973.

42. Zacharaki, E. I.; et al. Classification of Brain Tumor Type and Grade Using MRI Texture and Shape in a Machine Learning Scheme. *Magn. Reson. Med.* **2009,** *62*(6), 1609. DOI: 10.1002/MRM.22147.
43. Ganesan, M.; Sivakumar, N.; Thirumaran, M.; Saravanan, R. IoT and Cloud Based Brain Tumor Detection and Classification Model Using Optimal Densely Connected Convolutional Networks. *IIOABJ* [Online] **2020,** *11.* www.iioab.org (accessed Dec 22, 2022).
44. Sajjad, M.; Khan, S.; Muhammad, K.; Wu, W.; Ullah, A.; Baik, S. W. Multi-Grade Brain Tumor Classification Using Deep CNN With Extensive Data Augmentation. *J. Comput. Sci.* **2019,** *30,* 174–182. DOI: 10.1016/J.JOCS.2018.12.003.
45. Siddique, S.; Chow, J. C. L. Artificial Intelligence in Radiotherapy. *Reports Pract. Oncol. Radiother.* **2020,** *25*(4), 656–666. DOI: 10.1016/J.RPOR.2020.03.015.
46. Sherer, M. V.; et al. Metrics to Evaluate the Performance of Auto-Segmentation for Radiation Treatment Planning: A Critical Review. *Radiother. Oncol.* **2021,** *160,* 185–191. DOI: 10.1016/j.radonc.2021.05.003.
47. Chen, G.; et al. Rapid Progress in Intelligent Radiotherapy and Future Implementation. *Cancer Invest.* **2022,** *40*(5), 425–436. DOI: 10.1080/07357907.2022.2044842.
48. Tawalbeh, L.; Muheidat, F.; Tawalbeh, M.; Quwaider, M. IoT Privacy and Security: Challenges and Solutions. *Appl. Sci.* **2020,** *10*(12), 4102. DOI: 10.3390/APP10124102.

INDEX

H

I

L

M

N

O

P

Q

R

S